Catalysis Under
Transient Conditions

Catalysis Under Transient Conditions

Alexis T. Bell, EDITOR
University of California—Berkeley

L. Louis Hegedus, EDITOR
W. R. Grace & Company

Based on a symposium sponsored
by the Division of Colloid and
Surface Chemistry at the Second
Chemical Congress of the
North American Continent
(180th ACS National Meeting),
Las Vegas, Nevada,
August 27–28, 1980.

A C S   S Y M P O S I U M   S E R I E S **178**

AMERICAN CHEMICAL SOCIETY
WASHINGTON, D. C.   1982

Library of Congress CIP Data
Catalysis under transient conditions.
    (ACS symposium series, ISSN 0097–6156; 178)
    Includes bibliographies and index.
    1. Catalysis—Congresses.
    I. Bell, Alexis T., 1942-   . II. Hegedus, L. Louis,
1941-   . III. American Chemical Society. Division
of Colloid and Surface Chemistry. IV. Chemical Congress of the North American Continent (2nd: 1980:
Las Vegas, Nevada). V. Series.

QD505.C385   541.3'95   81–20639
ISBN 0–8412–0688–0 AACR2 ACSMC8 178 1–308
                                              1982

FOREWORD

The ACS SYMPOSIUM SERIES was founded in 1974 to provide a medium for publishing symposia quickly in book form. The format of the Series parallels that of the continuing ADVANCES IN CHEMISTRY SERIES except that in order to save time the papers are not typeset but are reproduced as they are submitted by the authors in camera-ready form. Papers are reviewed under the supervision of the Editors with the assistance of the Series Advisory Board and are selected to maintain the integrity of the symposia; however, verbatim reproductions of previously published papers are not accepted. Both reviews and reports of research are acceptable since symposia may embrace both types of presentation.

CONTENTS

PREFACE

The interest in the dynamic operation of heterogeneous catalytic systems is experiencing a renaissance. Attention to this area has been motivated by several factors: the availability of experimental techniques for monitoring species concentrations both in the gas phase and at the catalyst surface with a temporal resolution and sensitivity not previously possible, the development of efficient numerical methods for predicting the dynamics of complex reaction systems, and the recognition that in selected instances operation of a catalytic reactor under dynamic conditions can yield a better performance than operation under steady-state conditions.

The papers in this book were selected so that both the chemical and the engineering aspects of catalyst dynamics are represented; in addition, an attempt was made to draw a balance between theory and experiment.

The editors thank all the contributors to this volume and express their hope that the material will stimulate further interest in the dynamic operation of catalysts, both for its diagnostic powers and for its potential in improving reactor performance.

ALEXIS T. BELL
University of California
Berkeley, California

L. LOUIS HEGEDUS
W. R. Grace & Company
Columbia, Maryland

July 27, 1981

Understanding Heterogeneous Catalysis
Through the Transient Method

C. O. BENNETT

University of Connecticut, Department of Chemical Engineering, Storrs, CT 06268

Although catalysis can be understood through steady-state experiments, it seems clear that transient experiments will usually
furnish much additional information. Often steady-state data can
be explained by a number of different models, but the results of
transient experiments are usually so rich that only a detailed,
complex model will come close to explaining the results. These
ideas have long been applied in other fields, but in heterogeneous
catalysis they have come into acceptance only during the last 15
years or so. Pioneering experiments were done by Wagner (1) in
1938, and Tamaru (2) became an advocate of the method in 1963, em-
phasizing the measurement of adsorption during catalysis, the es-
sence of the transient method. A quantitative framework was laid
out in 1967 (3), and its implementation was carried out a few
years later by Kobayashi and Kobayashi (4) and by Yang et al. (5).
Effects due to transport resistances must be carefully avoided in work at atmospheric pressure, so that chemical effects having response times less than about a second cannot be followed.
These limitations are avoided at low pressure, and some of the
first transient results under vacuum conditions have been reported
by Bonzel and Ku (6) and by Jones et al. (7). Progress in transient studies up until about 1975 has been reviewed (8, 9).
The transient processes under discussion here must be described in terms of elementary steps, and the relative rates of
the steps will change as conditions change during transients.
This point of view has been advocated by Temkin (10), and it is
central to a recent discussion by Krylov (11). The adsorptive
capacity of the catalyst influences the transients--one has adsorption during reaction. Although modeling in terms of elementary steps is central to understanding catalysis and to modeling
reactors during transients, chemical engineers did not take this
idea into account until recently. From the first cycling studies
of Dorawala and Douglas (12) through the time of the review of
Schmitz in 1975 (13), reactor models were based on the use of a
steady-state rate expression in the component balances. This
state of affairs has now changed, and the importance of elementary
step models is generally acknowledged (14).

0097-6156/82/0178-0001$08.00/0

The bulk of this review concerns transient experiments on heterogeneous catalysis at atmospheric pressure. After some comments on current methods for doing the experiments, the application of the method will be illustrated by two examples: the oxidation of CO over Pt, and the hydrogenation of CO over Fe.

Some Experimental Methods

Gas-phase Kinetics. A better appreciation of the experiments to be discussed later will be obtained after a review of some experimental aspects of the transient method. Here we deal with experiments at atmospheric pressure. A flow sheet for kinetic measurements is given in Fig. 1, a descendant of that first given by Bennett et al. (15). Chemical analysis of the gases during transients is ideally done by a mass spectrometer, although Kobayashi and Kobayashi (4) used a number of gas chromatographs in order to get samples sufficiently frequently.

Step functions, pulses, and square waves can be generated with a low volume, chromatographic-type 4-way valve. We have found that the desired two gas mixtures are best made up and stored in cylinders rather than made continuously by blending two streams. At the time of the switch, there is a momentary stopping of the flow, and this usually results in a change in composition if the mixture is made by the continuous blending of two streams. By this method one or more spurious peaks are added to the desired step function. Naturally these are trivial for slow responses, but important for fast ones.

The needle valves of Fig. 1 are operated choked, so that there is a constant flow independent of downstream resistances. However, the valve shown upstream of the bubble meter is very important. It is adjusted so that the pressures at the switching valve and upstream are hardly perturbed by a switch. The flows of both mixtures are adjusted to be identical by the needle valves and the bubble meter. When all is properly regulated, the ball in each rotameter makes a short, quick (~1s) excursion at the moment of the switch, and identically for the return to the first position of the valve.

If the flow of each mixture to the switching valve is not identical, with minimal pressure disturbance at the switch, spurious peaks are also introduced. For instance in an experiment at steady state with H_2/CO (9/1) flowing to a differential reactor, a common transient experiment is a switch to H_2 alone, so that the hydrogen can react with a carbonaceous surface intermediate, produce methane, and effectively titrate the steady-state surface. If this experiment is not done by a switch from H_2/CO to H_2, but rather by suddenly closing the CO supply valve in a system using continuous gas blending, a peak of methane is formed which has nothing to do with titrating the surface. While H_2/CO is still in the line from the 4-way valve to the reactor (this

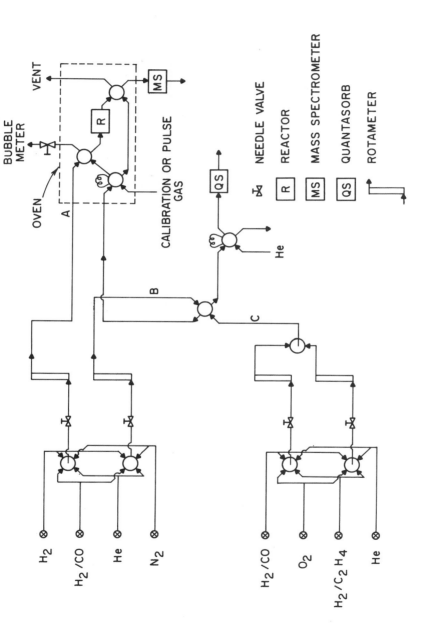

Figure 1. Flow sheet for transient experiments.

length should be minimized), its flow rate is reduced when the CO is cut off, conversion to CH_4 is increased, and there is a false peak at the beginning of the CH_4 transient caused by the reaction of H_2 with surface carbon.

The same effect is obtained by a switch from H_2/CO to helium flowing at a lower rate. There is a methane peak, leading to the erroneous idea that methane has been desorbed. With properly adjusted identical flows the CH_4 falls quickly to zero within the response time of the system – typically ~1s.

The four-way valve at the outlet of the reactor permits sending known gases to the mass spectrometer for calibration. The flow to the reactor need not be disturbed. This arrangement is particularly useful for measuring the water peak. It is well known that water adsorbs on the inside of the instrument. After a few days of pumping, preferably with bakeout, a stable water background peak can be obtained. A known water/helium mixture can be used to measure the peak caused by water entering through the continuous inlet. The response is slower than for other gases, but reproducible over several repitions a few minutes apart. However, if a reactor product stream containing water is sampled continuously, the background water peak gradually increases. Good analysis for water can be made by recording the 18–peak after a switch to dry helium via the reactor outlet 4–way valve. By thus keeping track of the varying background water peak we can make a good oxygen balance for the Fischer–Tropsch reaction at any instant; without the correct background correction this is not possible.

It is clear that the illustrated flowsheet is quite flexible. As shown, pulses can be produced by a 6–way valve. The valves are electrically actuated so a cycled feed can easily be produced. A separate mixture preparation apparatus has been built, so that preparing the various mixtures needed is quick and simple.

The 12" magnetic–sector mass spectrometer is computer–controlled. A typical arrangement is to jump the accelerating voltage over the desired mass numbers at a rate of about 5 numbers per second. The resulting voltages are appropriately converted to digital form, stored and used to give an output of mole fraction versus time for each component. Ordinarily hydrogen and helium cannot be obtained in such a moderately fast scan; for this the magnetic field would need to be changed. However, we have a separate side tube of small (4") radius, with a separate detector. With this arrangement a fast scan from 2 to about 100 is feasible. Either Faraday cup or electron multiplier detectors are used; the latter has the higher sensitivity but is subject to saturation effects.

A classical inlet system is used to obtain a fast response and to avoid mass fractionation. The gas to be sampled impinges on a capillary (150 μm I.D.) discharging into a small chamber (1mL) held at 10^2 Pa by an auxiliary pump. The viscous flow through the capillary is of the order of 5mL/mn. From the chamber a small sample of gas passes through *ca* 5–μm holes in a gold foil in Knudsen flow and through a tube leading to the spectrometer source.

Infrared Measurements. Elementary-step models can be fit to transient gas-phase data (5) which are obtained by the methods described above. The models will then predict the surface intermediate concentration during transients and at steady state. It is clear that it is also important to observe these surface species experimentally, during transients as well as at steady state. Infrared spectroscopy can be used during catalysis in the presence of the gas phase, so it plays an important role in transient studies.

Some of the first transient infrared data on heterogeneous catalysis were gathered by Heyne and Tomkins in 1966 (16). The method has been advocated and used by Tamaru, who has recently reviewed his work since 1967 in this field (17). Dent and Kokes used the transient method in IR studies in 1970 (18). Since then the method has been applied to kinetic studies in particular by Ueno et al. (19, 20) and most recently by Savatsky and Bell (21).

As already explained (9), the kinetic measurements are easier to interpret quantitatively if done either in a recirculating reactor (CSTR) or in a differential reactor but with measurable conversion. In the typical infrared cell we strive for a close approximation to a differential reactor. The conversion is usually too small to measure. Dwyer (22) and earlier Ueno et al. (19, 20) have used the optical system of Fig. 2. The cell of Fig. 3 permits cancelling gas phase adsorption bands. By operating the cell in the reflection mode as shown, positive control of the temperature of the catalyst disk is achieved, with a very low cell volume (~2 mL) and the possibility of cooling the cell window, as usually required by window material which is transparent over a wide range of frequencies. The temperature is measured in the metal block just behind the mirror. The thermal conductivity of the solid between the thermocouple and the surface is high compared to the effective conductivity of the gas which transfers heat by convection from the surface to the cooler window. Thus the surface temperature is assumed to be close to that measured by the thermocouple. Operation at 2 mL/s gives a residence time of 1s, and at this flow rate the 5 mg catalyst disk gives a negligible conversion for most systems.

Hegedus (23), for the study of adsorbed CO over a limited range of frequencies, was able to use sapphire windows which could be operated at 450°C. His reactor cell thus requires no cooling; it operates in transmission. The cell volume is about 22 mL. A flow rate of 250 mL/s was used so that the residence time was only 0.09s.

Since the flow system can be made with such a fast response, it becomes important to improve the response of the infrared detector. Use of a liquid-nitrogen-cooled Hg-CdTe detector, with 200 to 1000 Hz chopping and a lock-in amplifier, permits the IR

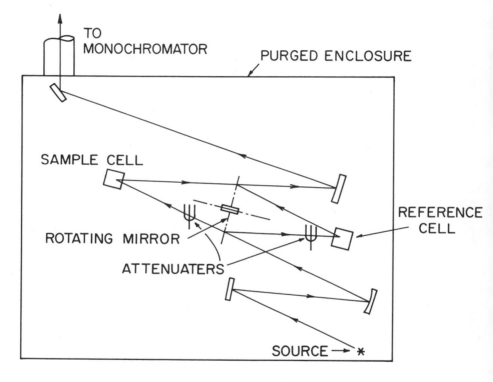

Figure 2.   IR optical system (19).

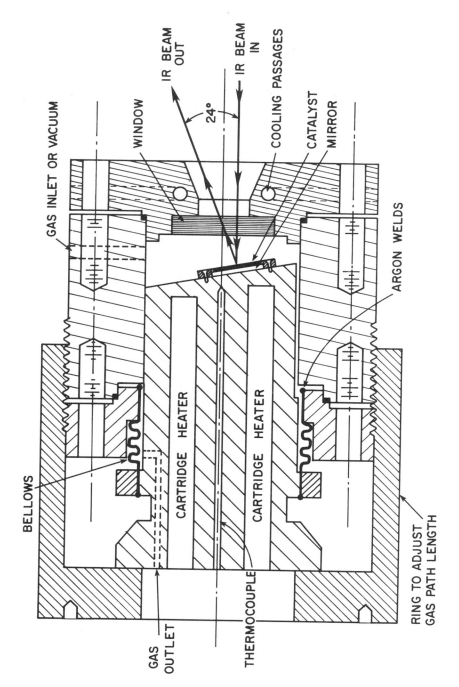

Figure 3.   IR cell reactor.

spectrometer to keep up with the imposed signal changes (23).
With a dispersion instrument the experiment can be run at several
constant wave numbers in repeated experiments. At the signal
levels often obtained in studies of the bands of adsorbed species,
a Fourier transform instrument may require about 20 scans of 1s
each to obtain an adequate interferogram, which of course represents only an average spectrum (21). Software has been developed
to resolve this problem through multiplexing in time and mirror
position (55, 56). However, this approach only works for a
repetitive transient, such as might result from a cycled feed.
During each scan the signal, mirror position, and time are stored.
The beginning of each scan is offset with reference to the begin-
ing of the transient by a time interval that may be as low as 50µs
(55). Then the program sorts the data accumulated over enough
repeated cycles for good signal to noise and reconstructs the
signal-averaged interferogram corresponding to the chosen time
intervals during the transient. This technique has been applied
to gas-phase transients (55, 56) but apparently not to surface
transients.

    With the dispersive IR instrument, it is important to use
computer control and data acquisition. For conventional experiments also, the attractiveness of such a system has been pointed
out by Peri (24).

    Measurements by Electron Spectroscopies. The various elec-
tron spectroscopies permit the determination of the surface
state of a solid because emitted electrons a few layers beneath
the surface do not escape. For example, emitted electrons may be
excited by a beam of incident electrons (Auger electron spectros-
copy-AES) or by a beam of X-rays (X-ray photoelectron spectros-
copy-XPS). The spectrum of intensity of the produced electrons is
recorded as a function of the energy of the electrons, leading to
a good estimate of surface composition. By XPS even the valence
state of a given element can often be determined. Usually XPS is
used in connection with transient studies only to measure the surface concentration of non-volatile intermediates like carbon (25);
the measurement must be made after the gas phase is removed. However, Matsushima et al. (26) have been able to follow by AES the
transient surface concentration of oxygen on platinum as the
oxygen reacts with carbon monoxide from the gas phase present at
working spectrometer pressures, 1.86×10^{-7} to 4.8×10^{-6} pascals.
The limitation of pressures to this range limits the number of
systems which can be observed during transients caused by changes
in the pressure (concentration) of the gas phase in contact with
the solid catalyst.

    Some progress has recently been made towards adopting XPS or
UPS to in situ measurements during transients. Rubloff (27) has
studied the interaction of CO with Ni(111) by directing a pulse
of CO (200-300 ms) to the nickel surface while observing the
$5\sigma/1\pi$ peak by UPS. The chamber pressure is little affected by

the pulse, but Rubloff estimates that the gas pressure near the
sample is about 150 times the system pressure during the pulse (27).
The pulse is achieved by a fast acting differentially pumped dosing
valve. A result is shown in Fig. 4.

Joyner and Roberts (28) have gone further by constructing a
sample cell, differentially pumped, within the spectrometer. Gas
is leaked at a constant rate into the sample region. There is a
small hole about 2 mm above the surface of the sample through
which the electrons excited from the sample pass into the spectrometer. X-rays are directed at the surface through an aluminum
window; a second small opening is provided if exitation is by UV
radiation. In this way the system pressure can be kept at 10^{-3} Pa
while there is a steady pressure of 100 Pa above the sample.
There is of course attenuation of the signal, as shown in Fig. 5,
taken from Joyner and Roberts (28). The gas phase spectrum will
also be obtained, but this usually can be separated easily from
the signal of the solid. This sample cell arrangement thus permits the study of the stationary-state surface during catalysis
and also its evolution in response to pulses and step functions in
the gas composition. The temperature of the sample should be controlled so that the surface can be studied during temperature-pro-
grammed desorption and reaction.

Modes of Operation

Steps and Pulses. Supposing that the proper equipment can be
devised for measuring gas-phase and--in some cases--surface transients, the question arises as to what kind of forcing function
should be used. An important class of experiments involves temperature as the perturbing variable. Temperature programmed desorption or reaction and flash desorption are widely used. Following transients induced by temperature change by mass spectrometric
analysis of the gas phase has been highly developed by Madix and
his students (29). In the present review we shall limit our discussion to concentration forcing functions.

Step functions up and down as perturbations have been related
to many common one- and two-step processes by Kobayashi and
Kobayashi (8). Figure 6, from our previous review (9) illustrates
the parametric sensitivity of this version of the transient method.
A switch is made from a flow of argon to N_2O/Ar over a NiO_2/SiO_2
catalyst at 350°C, and curve (1) of Fig. 6 shows the N_2 response.
This curve has been computer-simulated by a model which requires a
steady-state oxygen surface coverage θ of 0.58. Other values of θ
can be used in the simulation, and they can lead to the correct
steady-state rate, but as shown, the transient responses would be
unambiguously altered from that actually observed. Note that the
equilibrium oxygen surface coverage at the steady-state gas-phase
oxygen concentration would be lower than 0.58 (5).

All these experiments involve a multicomponent response,
ideally of both the gas phase and the surface. The number of

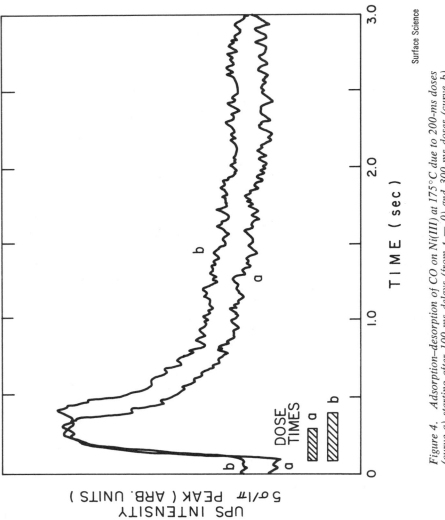

Figure 4. Adsorption–desorption of CO on Ni(III) at 175°C due to 200-ms doses (curve a) starting after 100-ms delays (from t = 0) and 300-ms doses (curve b) after similar delays (27).

Surface Science

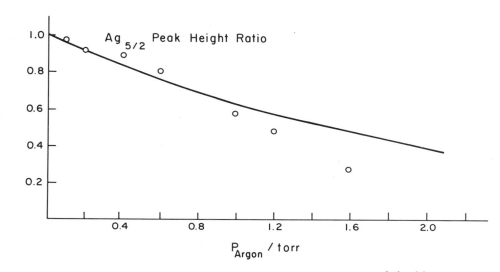

Surface Science

Figure 5.   The attenuation of Ag $3d_{5/2}$ signal by Ar. Key: ○, experimental points; and —, theory (28).

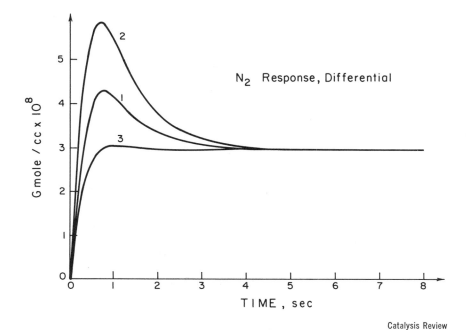

Catalysis Review

Figure 6.   Decomposition of N_2O over NiO in a differential reactor. The steady-state surface oxygen coverage θ was 0.58 (curve 1), 0.69 (curve 2), and 0.28 (curve 3) (9).

responses to fit--and thus the reliability of the result--can be increased by using stable isotopes. The quantitative manipulation of such a method has been most highly developed by Happel and co-workers (30). In this paper isotopic tracing is used to make it possible to identify the rate parameters of various steps in a sequence by a linear analysis. Methanation over a nickel catalyst has been studied. In the gas flowing to a gradientless reactor at steady state a step change is made from $H_2/^{12}CO$ to $H_2/^{13}CO$. The response is followed by mass spectrometry. All the chemistry, such as surface rates and coverages, is assumed to be unaltered by the change, so the result of the perturbation is that of a series of first-order processes, each one representing the accumulation of ^{13}C in a particular intermediate or gas component. We can speak of pools for the accumulation of ^{13}C without knowing the complete identity of each intermediate. By this method large concentrations of tracers can be used while retaining a convenient linear analysis. The mathematics of a similar superposition method has also been studied by Le Cardinal et al. (31). He and Happel have exchanged polemics on the matter (32, 33).

A disadvantage of tracer methods is the frequent presence of exchange processes which are difficult to account for. This hampered Happel's work with ^{18}O (30) and complicated the interpretation of the work of Conner and Bennett (34), who used pulses of tracers and a qualitative interpretation of their data on CO oxidation over NiO.

Cycled Feed. The qualitative interpretation of responses to steps and pulses is often possible, but the quantitative exploitation of the data requires the numerical integration of nonlinear differential equations incorporated into a program for the search for the best parameters. A sinusoidal variation of a feed component concentration around a steady state value can be analyzed by the well developed methods of linear analysis if the relative amplitudes of the responses are under about 0.1. The application of these ideas to a modulated molecular beam was developed by Jones et al. (7) in 1972. A number of simple sequences of linear steps produces frequency responses shown in Fig. 7 (7). Here ε is the ratio of product to reactant amplitude, η is the sticking probability, ω is the forcing frequency, and k_d is the desorption rate constant for the product. For the series process k_1 is the rate constant of the surface reaction, and for the branched process P is the fraction reacting through path 1 and desorbing with a rate constant k_{d1}. This method has recently been applied to the decomposition of hydrazine on Ir(111) by Merrill and Sawin (35).

Whether to use as perturbing function a step, pulse, or cycled feed depends on the particular system under study. For expensive tracers, a pulse is often mandatory. However, simple textbook relations based on a Dirac function do not usually apply, for a relatively long pulse may be required to get a good signal. A long enough pulse becomes two step functions, and as already men-

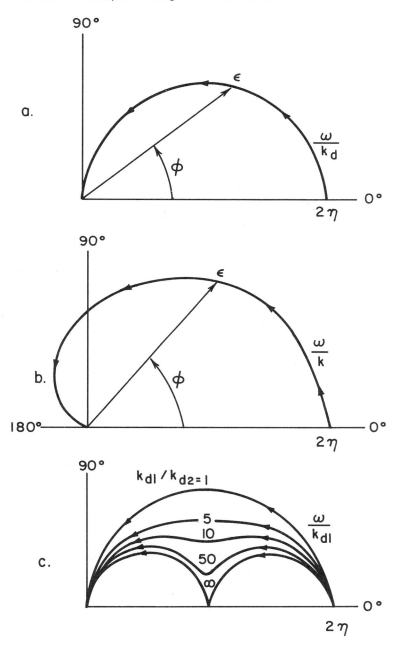

Figure 7.   Polar plots of the reaction product vector for three surface mechanisms. Key: a, simple adsorption–desorption; b, series process, $k_1 = k_d = k$; and c, branch process, $P = 0.5$ (7).

tioned, there is a convenient directory of common responses (8).
For a cycled feed (a chain of interfering pulses) the interpretation requires some mathematical transformations, but results such
as Fig. 7 can also be used as qualitative indicators of certain
processes (35). The complete computer simulation is quite tedious
and has so far been done for relatively few reactions; thus the
partial exploitation of the data in the spirit of preceding work
is often all that is attempted in practice.

Strozier (36) has also used a cycled (pulsed) feed at UHV.
As in the molecular beam experiment, the reactor volume, pumping
speed, and rate of introduction of reactants have values which lead
to a flux of reactants well defined in time. Strozier, however,
simply doses gas into the vacuum system (reactor) rather than using
a molecular beam. He studied CO oxidation, which has nonlineari-
ties in the surface rate equation, so that computer rather than
analytic solutions are necessary. The results are represented at
constant frequency and varying temperature as shown in Fig. 8,
which is a computer simulation (37).

Cycled-feed experiments are more complicated to analyze if
done at atmospheric pressure and at relatively high conversion.
Now the surface reaction rates alter the gas-phase reactant concentrations. Cutlip (38) has studied CO oxidation over Pt/Al_2O_3
in a gradientless reactor under conditions often leading to complete conversion. The feed gas alternated between 2% CO and 3% O_2
in argon. Figure 9 shows some typical results. Clearly there is
no hope of simulating such data by anthing but a complicated computer model.

Figure 10 (38) shows that under some circumstances the rate
is greatly enhanced (20 times) by a cycled feed. At the conditions shown there are no multiple steady states, the only steady
state corresponds to a very high CO surface coverage. By alternately switching to pure oxygen, CO_2 is generated at a high rate
from the adsorbed CO. The net result is as if the surface were
more equally covered with oxygen and carbon monoxide with a
Langmuir-Hinshelwood rate equation: $r = k (CO_{ads})(O_{ads})$.

Figure 11 shows the effect on CO conversion (oxidation) over
two Rh/Al_2O_3 catalysts of cycling at 550°C as studied by Schlatter
and Mitchell (39). Steady state operation (high frequency) gives
complete conversion of a stoichiometric reactant mixture, and
cycling around this steady state ratio must give a lower average
conversion. Since the control operation to maintain an average
stoichiometric ratio in the reactor feed leads to cycles of frequency about 1 Hz (39), it is of practical interest to understand
these effects. Cerium has been added to the catalyst to try to
retain oxygen and keep the rate up during the part of the cycle
during which there is insufficient gaseous oxygen for complete
conversion. As shown, Schlatter and Mitchell (39) found the Ce-
containing catalyst superior, but the explanation for its behavior
is more complicated than simple oxygen storage.

The possible importance of obtaining improved yield in a

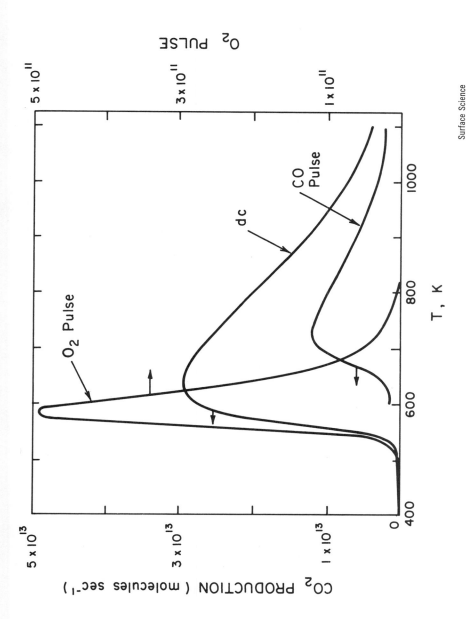

Figure 8.   Responses averaged in time to steady and pulsed feeds for CO oxidation over Pt (37).

Surface Science

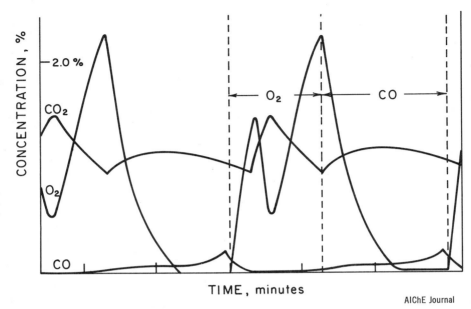

Figure 9. Reactor outlet concentrations during a 3-min period at 60°C (38). Feed is cycled from 2% CO at 80 s to 3% O_2 at 100 s over a Pt/Al_2O_3 catalyst at 60°C.

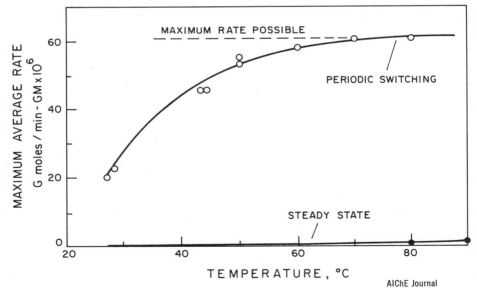

Figure 10. Rate comparison of periodic to steady-state operation for stoichiometric feed of CO/O_2 to Pt/Al_2O_3 catalyst (38).

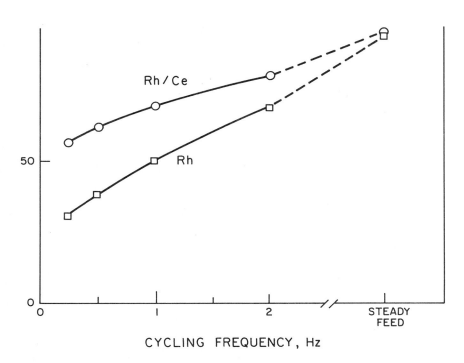

Figure 11.   Effect of Ce on the cycled performance of a Rh/Al_2O_3 catalyst fed CO/O_2 mixtures at 550°C and a space velocity of 104,000 h^{-1} (39).

multiproduct catalytic process by feed concentration cycling has
already been discussed in our previous review (9). Bilimoria and
Bailey (40) have studied the hydrogenation of acetylene to ethylene
and ethane, and Al-Taie and Kerschenbaum (41) have studied the hydrogenation of butadiene to butylene and to butane. In the latter
work the yield of butylene was improved by cycling for certain
reactor conditions. No mechanistic analysis is done in these
papers from the Chemical Reaction Engineering Meeting in Houston
beyond acknowledging that consideration of surface rates and
capacities is essential.

    Self-sustained Oscillations. Under certain conditions, iso-
thermal limit cycles in gaseous concentrations over catalysts are
observed. These are probably caused by interaction of steps on
the surface. Sometimes heat and mass transfer effects intervene,
leading to temperature oscillations also. Since this subject has
recently been reviewed (42, 43) only a few recent papers will be
mentioned here.

    Cutlip and Kenney (44) have observed isothermal limit cycles
in the oxidation of CO over 0.5% Pt/Al_2O_3 in a gradientless reactor
only in the presence of added 1-butene. Without butene there were
no oscillations although regions of multiple steady states exist.
Dwyer (22) has followed the surface CO infrared adsorption band
and found that it was in phase with the gas-phase concentration.
Kurtanjek et al. (45) have studied hydrogen oxidation over Ni and
have also taken the logical step of following the surface concentration. Contact potential difference was used to follow the oxidation state of the nickel surface. Under some conditions,
oscillations were observed on the surface when none were detected
in the gas phase. Recently, Sheintuch (46) has made additional
studies of CO oxidation over Pt foil.

    The N_2O decomposition, CO oxidation, and H_2 oxidation reac-
tions are known to exhibit concentration oscillations over noble
metal catalysts. Flytzani-Stephanopoulos et al. (47) have observed oscillations for the oxidation of NH_3 over Pt. The effects
are dramatic and lead to large temperature cycles for the catalyst
wire. Heat and mass transfer effects are important.

The Oxidation of Carbon Monoxide on Platinum

    This subject has been thoroughly reviewed by Engel and Ertl
(48) and most recently by White (49). Here the intention is to
describe some contributions of the transient method to our understanding of this classical problem. Only the temperature range
below 450K is discussed here.

    The question of the role of the Eley-Rideal (ER) mechanism in
the reaction is usually investigated by the reaction of preadsorbed oxygen with CO from the gas phase. The well known results
of Bonzel and Ku (6) are shown in Fig. 12. CO_2 is immediately
produced, lending some support to the existance of an ER step.
However, Matsushima et al. (26) have used Auger electron spectros-

copy to follow the surface concentration of oxygen (dissociated
into adsorbed atoms) during the same process. Although the rate
is about first-order in gas-phase CO concentration, it is zero-
order in surface O, as shown in Fig. 13. The order does increase
as the surface oxygen is depleted. The result is interpreted as
showing that CO does not react by impact on oxygen atoms but must
adsorb in a mobil precursor state.

In a later article, Matsushima (50) has used a transient
tracer method at 301K to study the ER process further. The surface
of the Pt was initially exposed to a pulse of a mixture of $^{13}CO/$
^{12}CO so that the CO surface coverage was 0.23 and the ratio of
$^{13}CO(a)$ to $^{12}CO(a)$ was 1.6. Then a pulse of $^{12}CO + O_2$ was admit-
ted, and the first CO_2 to be produced had the same $^{13}C/^{12}C$ ratio
as the preadsorbed CO. This result is interpreted as meaning that
the surface oxygen reacts with surface CO much faster than with
the gaseous CO; the ER step plays a minor role.

This process has been studied by infrared spectroscopy by
Dwyer (22), using 9% Pt/SiO_2 and the cell used by Ueno et al. (19).
The bands arising from gaseous CO_2 and surface CO were followed
after a switch in the gas phase composition from 3kPa O_2 in helium,
to pure helium for a few seconds and then to 2kPa CO in helium.
This is basically the same sequence as that of Fig. 12. The results are shown in Fig. 14. The CO(a) band grows immediately and
at the same time CO_2 is produced. Taken with the results of
Matsushima et al. (26, 50) it seems reasonable that the ER step
plays a minor role in the reaction between CO and O(a).

The process $O_2 + CO(a)$ is also interesting and this can be
studied by a gas phase switch from CO to O_2. As mentioned above
a short exposure to helium between the CO and O_2 is used. This
procedure eliminates the interaction of CO and O_2 during the mixing time (ca 2 s) for the cell. Oxygen does not desorb in helium
(T<450K) but Dwyer (22) found that CO does, as shown in Fig. 15.
The process was studied so that suitable corrections could be made
for the CO → He → O_2 experiments. The desorption of Fig. 15 starts
from a surface coverage of CO near unity. From the initial rates
an activation energy of 8 kcal/mole is obtained. This may be compared with 9 kcal/mole reported by Nashayima and Wise (51). Figure 16 (22) shows that the slight desorption of loosely bound CO
has a remarkably large effect on the CO(a) left on the surface.

The results of the CO → He → O_2 experiment (22) are shown in
Fig. 17. There is an induction time between the admission of O_2
and start of the reaction of CO(a). Oxygen cannot react from the
gas phase, and its adsorption must await the desorption of a trace
of CO(a). Bonzel and Ku () found a similar result, as shown in
Fig. 18. Their induction times are much longer than those of
Dwyer (22), perhaps because of the regularity of their (110) surface.

From these examples it is clear that the transient results
add important understanding to our ideas about the oxidation of CO
over Pt.

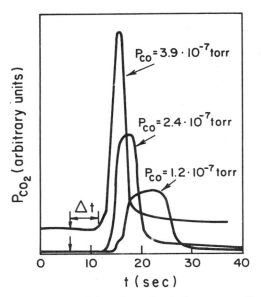

Figure 12.   Reaction between CO and preadsorbed oxygen on Pt (110) at 218°C (6).

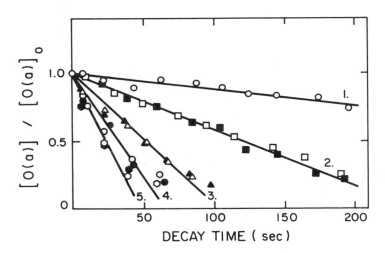

Figure 13.   Reaction of surface O with CO for several fixed CO pressures over Pt measured by Auger electron spectroscopy. Key: Fixed CO pressures (Pa) are 1.86 × 10⁻⁷ (1), 6.27 × 10⁻⁷ (2), 1.67 × 10⁻⁶ (3), 2.8 × 10⁻⁶ (4), and 4.8 × 10⁻⁶ (5). The light and dark symbols denote separate experiments (26).

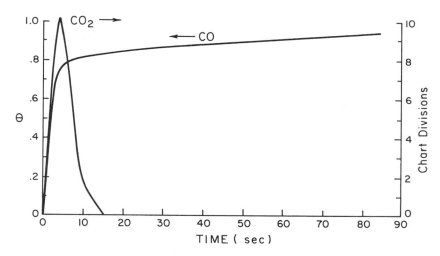

Figure 14.   Reaction between CO and O_2 preadsorbed on Pt/SiO_2 (22) at 160°C and flow rate of 9.5 mL/60 s.

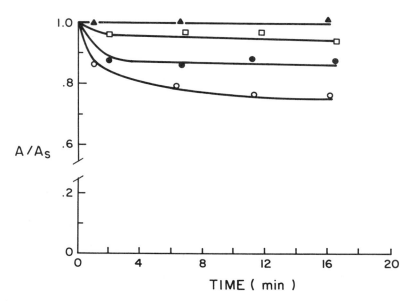

Figure 15.   Desorption of CO from Pt/SiO_2 (22). Key: ▲, 100°C; □, 120°C; ●, 140°C; and ○, 160°C.

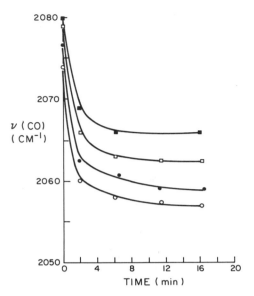

Figure 16.   Variation of IR frequency during CO desorption from Pt/SiO₂. Key:
■, 100°C; □, 120°C; ●, 140°C; and ○, 160°C (22).

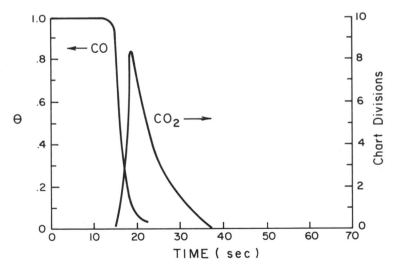

Figure 17.   Reaction between O₂ and CO preadsorbed on Pt/SiO₂ (22) at 160°C
and flow rate of 9.5 mL/60 s.

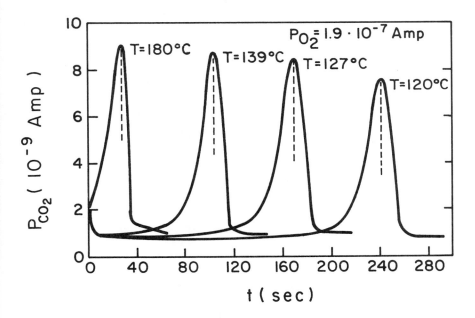

Figure 18.   Rate of CO_2 formation on Pt (110). Initially, the Pt (110) surface was exposed to a feed mixture of CO and oxygen. Then at $t = 0$, the partial pressure of CO was suddenly reduced and the subsequent rate of CO_2 formation was measured (6).

The Reaction of Hydrogen and Carbon Monoxide over Iron

The transient method has contributed to our understanding of H_2/CO reactions, and a few results of studies over iron will show the power of the method. Matsumoto and Bennett (52) worked with a CCI fused iron catalyst (ca 96% iron) which had been reduced in flowing H_2 at 450°C for at least 15 h. Upon switching to a feed of 10% CO in H_2, the data of Fig. 19 were obtained. Large rates of production of H_2O and CO_2 are observed as CO is dissociated on the fresh catalyst. The production of methane and higher hydrocarbons starts at zero and gradually increases. This result was interpreted as evidence that surface carbon reacted preferentially with iron to form a carbide. As this reaction slowed with time, surface carbon built up and led to the production of methane and lower production rates for CO_2 and H_2O. Figure 20, from Reymond et al. (25) shows the evolution of the methane production rate over a longer period. A switch from the H_2/CO feed to pure H_2 produced a peak of methane as shown (250°C). This increase in CH_4 production rate shows that H_2 must be adsorbed (probably dissociated as H) in order to react, and also that the steady-state surface must be largely covered with carbonaceous intermediate. Methane production seems to be governed by the equation $r = k\ C(a)\ H(a)$, with $C(a) > H(a)$ so that an increase in $H(a)$ leads to a higher rate. The second peak in Fig. 20 represents methane generated by the decarburization of the bulk of the iron.

When the freshly reduced catalyst is suddenly exposed to 10% C_2H_4 in H_2 the result is shown in Fig. 21 (25). Chain growth occurs as with CO/H_2, but the reaction rate decreases from an initial maximum as the surface becomes covered with intermediates. A switch to pure H_2 produces only a trace of methane, even on heating to 500°C, so no surface carbon or bulk carbide has been formed. Probably the sequence of steps passes through a CH_2 intermediate, whereas C is needed for carburizing the iron. The rate of reaction of C_2H_4/H_2 is higher than that of CO/H_2, lending weight to the concept of the hydrogenation of $C(a)$ as the rate-determining step for the latter.

Some of these same experiments have been done using 10% Fe/Al_2O_3 rather than the fused iron catalyst (53). Figure 22 shows the result of a switch from H_2 to 10% CO in H_2 over a freshly reduced catalyst. Here a large initial rate of methane formation is observed and water does not appear until most of the initial peak has passed. The probable explanation for the presence of the CH_4 peak is that water produced by methanation is adsorbed on the initially dry γ-Al_2O_3 support (100 m^2/g). Thus the iron remains briefly in a relatively reduced state. For the CCI catalyst the Al_2O_3 promoter is not sufficient to prevent the water from rising quickly as shown in Fig. 19. The H/O ratio on the surface is reduced, and carburization occurs more rapidly than methanation, as for the unsupported catalyst.

When the 10% Fe/Al_2O_3 catalyst is pretreated with water vapor,

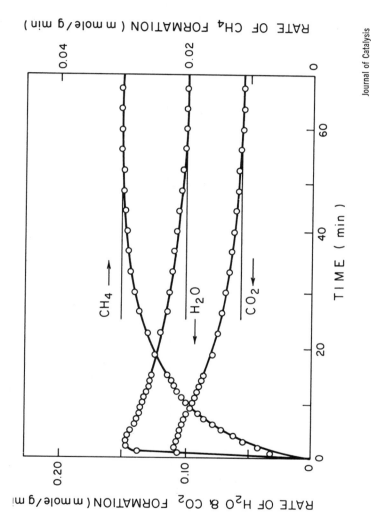

RATE OF CH₄ FORMATION (m mole/g min)

RATE OF H₂O & CO₂ FORMATION (m mole/g mi

TIME (min)

Figure 19. Formation of major components as a function of exposure time for the reaction of 10% CO/H₂ at 250°C over CCI fused iron catalyst (reduced at 450°C) (52).

Journal of Catalysis

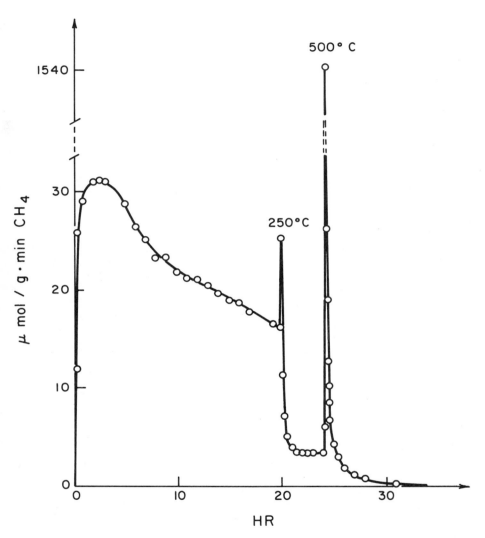

Figure 20.   Hydrogenation of surface carbon over catalyst initially reduced for 15 h at 500°C and then cooled to 250°C. At t = 0 flow is switched to H_2/CO = 9, 250°C. At 20 h feed is switched to pure H_2, 250°C; at 24 h temperature is raised to 500°C (25).

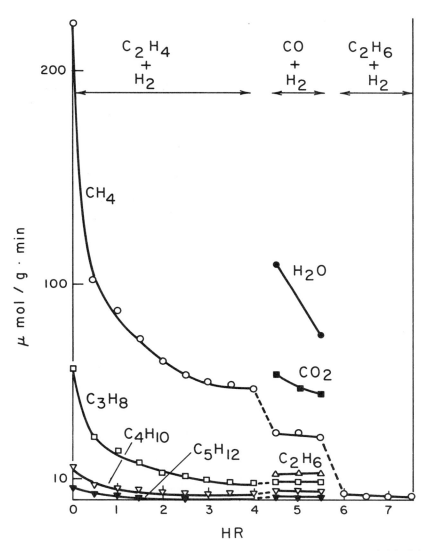

Figure 21.   Reaction of ethylene and hydrogen (1:10) on a reduced catalyst at 250°C, followed by switch to CO/H₂ (1:10) and then C₂H₆/H₂ (1:10) (25).

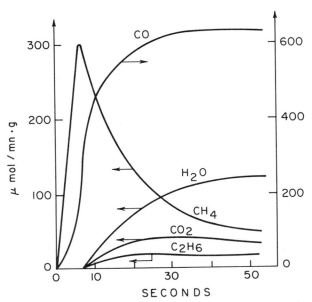

Figure 22.   Initial production rates of principal products from a 10% Fe/Al₂O₃
catalyst (53).

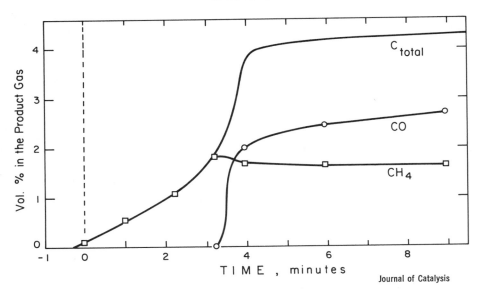

Figure 23.   Response of the 10% Ni/SiO₂ catalyst to a step increase in P_{CO} (54).
Conditions: 190°C; P_{H_2} 1 atm; t < 0, P_{CO} = 0; t > 0, P_{CO} = 0.44 atm; and flow
rate = 100 mL/60 s.

the methane production behaves as for the CCI catalyst--it rises
from zero as in Fig. 19.

Van Ho and Harriott (54) have done similar transient experiments on H_2/CO over 10% Ni/SiO_2, and a result is shown in Fig. 23.
Even though the bulk of the nickel is not carbided, the methane
production rate rises from zero, as shown.

The complete discussion of these experiments is beyond the
scope of this review. By showing these few results some idea of
the richness in information of the transient data is conveyed.
Any model which can explain all such complicated results is bound
to be nearer the truth than one based on steady state data alone.
An understanding improves, it is permitted to hope that this will
eventually lead to the development of better catalysts.

Summary

From the examples discussed in this review it is clear that a
proper understanding of heterogeneous catalysis is linked to understanding transient effects. The interpretation requires a model
based on elementary steps, and various steps may be rate-determining as the conversion and temperature vary. Complicated programs
for fitting a steady-state equation to data are not sufficient if
the equation itself changes with conversion, for example. Any
mathematical control or design calculations in reactor design must
take into account the elementary steps and the storage capacity of
the catalyst surface.

Understanding catalysis has been emphasized here, but the
possible practical effects of deliberate cycling of a reactor feed
to improve yield must also be kept in mind. Also, undesired feed
oscillations must be taken into account in the design of a reactor.
Design for such a situation is not possible based on steady state
laboratory data alone.

Acknowledgement

This work was supported by NSF grants #ENG-7921890, and #CPE-
7901018 A01.

Literature Cited

1. Wagner, Carl, and Hauffe, Karl. Untersuchungen über den
    stationären Zustand von Katalysatoren bei heterogenen
    Reaktionen. I: Ztschr. Elektrochem. 1938, 44 (3) 172;
    II: ibid. 1939, 45 (5) 409.
2. Tamaru, Kenzi. Adsorption measurements during surface cataly-
    sis. Adv. Catal. 1964, 15, 65.
3. Bennett, C. O. A dynamic method for the study of heterogene-
    ous catalytic kinetics. A.I.Ch.E. J. 1967, 13, 890.
4. Kobayashi, M., and Kobayashi H. Application of transient re-
    sponse method to the study of heterogeneous catalysis. J.
    Catal. 1972, 27, 100, 108, 114.

5. Yang, C. C., Cutlip, M. B., and Bennett, C. O. A study of nitrous oxide decomposition on nickel oxide by a dynamic method. Proc. of the Vth Congr. on Catal., Palm Beach, 1973.

6. Bonzel, H. P., and Ku, R. Mechanisms of the catalytic carbon monoxide oxidation on Pt(110). Surface Sci. 1972, 33, 91.

7. Jones, R. H., Olander, D. R., Siekhaus, W. J., and Schwarz, J.A. Investigation of gas-solid reactions by modulated molecular beam mass spectrometry. J. Vac. Sci. Technol. 1972, 9, 1429.

8. Kobayashi, M., and Kobayashi, M. Transient response method in heterogeneous catalysis. Catal. Rev.-Sci. Eng. 1974, 10 (2), 139.

9. Bennett, C. O. The transient method and elementary steps in heterogeneous catalysis. Catal. Rev.-Sci. Eng. 1976, 13 (2), 121.

10. Temkin, M. I. Relaxation rate of a two-stage catalytic reaction. Kin. and Catal. 1976, 17 (5), 1095.

11. Krylov, O. V. Elementary processes determining the mechanisms of a catalytic reaction. Kin. and Catal. 1980, 21 (1), 79.

12. Dorawala, T. G., and Douglas, J. M. Complex reactions in oscillating reactors. A.I.Ch.E. J.. 1971, 17 (4), 974.

13. Schmitz, Roger A. Multiplicity, stability, and sensitivity of states in chemically reacting systems--A review. Adv. Chem. Sci. 1975, 148, 154.

14. Sheintuch, M., and Schmitz, R. A. Kinetic modeling of oscillatory catalytic reactions. ACS Symp. Ser. 1978, 65, 487.

15. Bennett, C. O., Cutlip, M. B., and Yang, C. C. Gradientless reactors and transient methods in heterogeneous catalysis. Chem. Eng. Sci. 1972, 27, 2255.

16. Heyne, H., and Tomkins, F. C. Application of infrared spectroscopy and surface potential measurements in a study of the oxidation of carbon monoxide on platinum. Proc. Roy. Soc., Ser. A. 1966. 292, 460.

17. Tamaru, K. Recent progress in elucidating the mechanism of heterogeneous catalysis. Pure and Appl. Chem. 1980. 52, 2067.

18. Dent, A. L., and Kokes, R. J. Intermediates in ethylene hydrogenation over zinc oxide. J. Phy. Chem. 1970. 74, 3653.

19. Ueno, A., Hochmuth, J. K., and Bennett, C. O. Interaction of CO_2, CO, and NiO studied by infrared spectroscopy. J. Catal. 1977. 49, 225.

20. Ueno, A., and Bennett, C. O. Infrared study of CO_2 adsorption on SiO_2, J. Catal. 1978. 54, 31.

21. Savatsky, B. J., and Bell, A. T. Studies of NO reduction by H_2 using transient response techniques. This volume.

22. Dwyer, S. M. Transient infrared studies of carbon monoxide oxidation on a supported platinum catalyst. M.S. Thesis, University of Connecticut, 1980.

23. Hegedus, L. L. Infrared spectroscopic reactor system to study the transient operation of heterogeneous catalysts.

24. Peri, J. B. Characterization of catalyst surfaces by computerized infrared spectroscopy. Prepr., Div. Pet. Chem., Am. Chem. Soc. 1978. 23 (4), 1281.
25. Reymond, J. P., Mériaudeau, P., Pommier, B., and Bennett, C. O. Further results on the reaction of H_2/CO on fused iron by the transient method. J. Catal. 1980. 64, 163.
26. Matsushima, T., Almy, D. B., and White, J. M. The reactivity and Auger chemical shift of oxygen adsorbed on platinum. Surface Sci. 1977. 67, 89-108.
27. Rubloff, G. W. Photoemission studies of time-resolved surface reactions: Isothermal desorption of CO from Ni (111). Surface Sci. 1979. 89, 566.
28. Joyner, R. W., and Roberts, M. W. A high-pressure electron spectrometer for surface studies. Surface Sci. 1979. 87, 501.
29. Johnson, S. W., and Madix, R. J. Modification of adsorption and desorption of CO and H_2 and surface reactions of alcohols by sulfur on Ni (100). ACS Symp. Ser. 1981; Madix, R. J. Adv. in Catal. 1980. 29.
30. Happel, J., Suzuki, I., Kokayeff, P., and Fthenakis, V. Multiple isotope tracing of methanation over nickel catalyst. J. Catal. 1980. 65, 59.
31. Le Cardinal, G., Walter, E., Bertrand, P., Zoulalian, A. and Gelas, M. A new operating method for the kinetic study of open systems by means of tracer elements. Chem. Eng. Sci. 1977. 32, 733.
32. Happel, J. Transient tracing. Chem. Eng. Sci. 1978. 33, 1567.
33. Le Cardinal, G. Chem. Eng. Sci. 1978. 33, 1568.
34. Conner, W. C., and Bennett, C. O. Carbon monoxide oxidation on nickel oxide. J. Catal. 1976. 41, 30.
35. Merrill, R. P., and Sawin, H. H. Decomposition of hydrazine on Ir (111). This volume.
36. Strozier, J. A., Jr. Oxidation of CO on Pt by an a.c. pulsing technique. II. Experiment. Surface Sci. 1979. 87, 161.
37. Strozier, J. A., Jr. Cosgrove, G. J., and Fischer, D. A. Oxidation of CO on Pt and Pd. I. Theory. Surface Sci. 1979. 82, 481.
38. Cutlip, M. B. Concentration forcing of catalytic surface rate processes. A.I.Ch.E. J. 1979. 25 (3) 502.
39. Schlatter, J. C., and Mitchell, P. J. Three-way catalyst response to transients. Ind. Eng. Chem. Prod. Res. Dev. 1980. 19, 288.
40. Bilimoria, M. R., and Bailey, J. E. Dynamic studies of acetylene hydrogenation on nickel catalysts. ACS Symp. Ser. 1978. 65, 43.
41. Al-Tai, A. S., and Kerschenbaum, L. S. Effect of periodic operation on the selectivity of catalytic reactions. ACS Symp. Ser. 1978. 65, 42.

42. Sheintuch, M., and Schmitz, R. A. Oscillations in catalytic
     reactions. Catal. Rev. 1977. 15 (1) 107.
43. Slin'ko, M. G., and Slin'ko, M. M. Self-oscillations of
     heterogeneous catalytic reaction rates. Catal. Rev.-Sci. Eng.
     1978. 17 (1) 119.
44. Cutlip, M. B., and Kenney, C. N. Limit cycle phenomena during
     catalytic oxidation reactions over a supported platinum cata-
     lyst. ACS Symp. Ser. 1978. 65, 39.
45. Kurtanjek, A., Sheintuch, M., and Luss, D. Surface state and
     kinetic oscillations in the oxidation of hydrogen on nickel.
     J. Catal. 1980. 66, 11.
46. Sheintuch, M. Oscillatory state in the oxidation of carbon
     monoxide on platinum. A.I.Ch.E. J. 1981. 27, 20.
47. Flytzani-Stephanopoulos, M., Schmidt, L. D., and Caretta, R.
     Steady state and transient oscillations in NH_3 oxidation on
     Pt. J. Catal. 1980. 64, 346.
48. Engle, T., and Ertl, G. Elementary steps in the catalytic
     oxidation of carbon monoxide on platinum metals. Advances
     in Catalysis. 28. New York: Academic Press, Inc., New
     York. 1979.
49. White, J. M. Transient low-pressure studies of catalytic
     carbon monoxide oxidation. This volume.
50. Matsushima, T. Kinetic studies on the CO oxidation over
     platinum by means of carbon 13 tracer. Surface Sci. 1979.
     79, 63.
51. Nishiyama, Y., and Wise, H. Surface interactions between
     chemisorbed species on platinum: Carbon monoxide, hydrogen,
     oxygen, and methanol. J. Catal. 1974. 32, 50.
52. Matsumoto, H., and Bennett, C. O. The transient method
     applied to the methanation and Fischer-Tropsch reactions over
     a fused iron catalyst. J. Catal. 1978. 53,331.
53. Teule-Gay, F. Transient studies of the CO/H_2 reaction on
     various supported iron catalysts. M. S. Thesis, University
     of Connecticut, 1981.
54. Van Ho, S., and Harriott, P. The kinetics of methanation on
     nickel catalysts. J. Catal. 1980. 64, 272.
55. Murphy, R. E., Cook, F. H., and Saki, H. J. Opt. Soc. Am.
     1975. 65, 600.
56. Mantz, A. W. Appl. Spectroscopy. 1976. 30, 459.

RECEIVED July 28, 1981.

Transient Low-Pressure Studies of Catalytic Carbon Monoxide Oxidation

J. R. CREIGHTON and J. M. WHITE

University of Texas, Department of Chemistry, Austin, TX 78712

The purpose of this article is to review the results of transient low pressure studies of carbon monoxide oxidation over transition metal substrates. Particular emphasis is given to the use of in-situ electron spectroscopy, flash desorption, modulated beam and titration techniques. The strengths and weaknesses of these will be assessed with regard to kinetic insight and quantification. An attempt will be made to identify questions that are ripe for investigation. Although not limited to it, the presentation emphasizes our own work. A very recent review of the carbon monoxide oxidation reaction (1) will be useful to readers who are interested in a more comprehensive view.

The carbon monoxide oxidation reaction has been widely studied over the course of many years. The pioneering work of Langmuir (2) set the tone for much of the subsequent work and we continue to refine many of his ideas. Catalyzed carbon monoxide oxidation is perhaps better understood than any other heterogeneous reaction and, because it shows little variation with crystal plane (3), there is a good correlation of low area single crystal results with high area supported catalyst results. This is particularly true under conditions where CO accumulates and inhibits the rate of CO_2 production (4,5). In the realm of applications, CO oxidation is a significant consideration in systems designed to limit undesirable emissions from combustion processes.

At the outset, it is worth noting how transient data fit into the overall picture of this reaction and what we expect to learn by doing such experiments. With the exception of oscillations (6), transients are induced by rapidly changing one or more of the independent variables of the system. For the case

of CO oxidation, changing the substrate temperature
or the partial pressures of CO or O_2 induces such
transients. The information generated comes from
measurement of the time evolution of surface coverages
and gas phase composition. The former are, of course,
central to any kinetic description involving adsorbed
species; measurements under working conditions give
the greatest insight because they are directly
related to the CO_2 production rates. The latter
provide product formation and adsorption/desorption
rates during transients. In addition, when used in
conjunction with a model, coverages can be calculated
from pressure versus time data. Taken together these
measurements in principle provide the rates and
coverages needed for a full kinetic description. In
practice, there are serious experimental limits on
the measurements. In particular the dynamic range
over which surface concentrations can be reliably
measured is 0.01 to 1 monolayer; the lower end of
this range is particularly troublesome. Moreover,
because adsorbed species often occupy sites of
variable activity on a surface, measurement of the
total concentration of an adsorbed species may be
insufficient. Improved resolution and measurement
of small concentrations of relatively labile surface
species lie at the heart of experimental progress
in the surface chemistry of heterogeneous catalysis.
Succinctly put, the old catalytic chemists' proverb,
"if you can see a certain surface species, it is
not kinetically important", must be taken seriously.
However, in the case of CO oxidation, the kinetically
active adsorbed oxygen and carbon monoxide species
can in fact be measured under many, but not all,
conditions.

Transients Induced by Temperature Changes

One of the standard surface science methods for
assessing the concentration and stability of a
chemisorbed species is thermal desorption spectroscopy
(TDS). An early paper by Redhead (7) developed the
conceptual framework for certain cases. Many papers
since then have expanded the applicability of this
method. Recent work of Madix (8), Weinberg (9) and
Schmidt (10) is particularly noteworthy. Most of
this work focuses on the desorption of a single
molecular species and not on reactions in desorbing
systems. However, qualitative features of the
temperature dependence of reactions can be assessed
using this method. Figures 1 and 2 taken from the

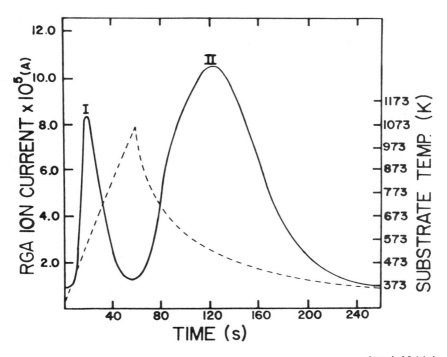

Journal of Catalysis

Figure 1.   Variation of CO_2 pressure with time during the flash of a Pd substrate preexposed to oxygen at high temperatures.   The total pressure was about 10^{-4} Pa and $O_2/CO \simeq 1.5$.   Key: – – –, temperature variation; and ——, CO_2 pressure (11).

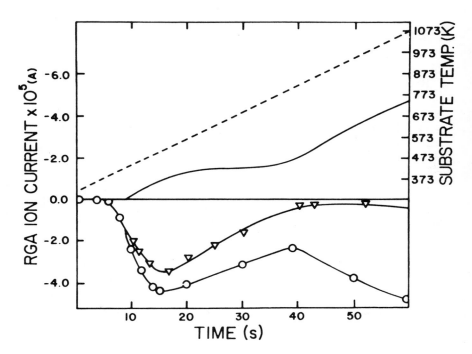

Figure 2.   Non-steady state variation of O_2 pressure with time during flashing of a stable polycrystalline Pd substrate. Initial $O_2/CO = 1.5$. The origin of the ordinate is defined as the O_2 pressure at 300 K. Key: ○, observed O_2; ▽, calculated on the basis of CO_2 production; ——— , difference between calculated and observed; and — — — , temperature (11).

work of John Close (11) in our laboratory adequately
illustrate the point.

Figure 1 shows the variation of the CO_2 pressure
during the flash and subsequent cool down of a polycrystalline Pd foil. The sample, mounted within a
continuously pumped UHV system, was exposed to 10^{-6}
torr of a mixture characterized by $O_2/CO = 1.5$. After
stabilization at 300K, the sample was heated resistive-
ly at a rate of $12Ks^{-1}$ while the CO_2 partial pressure
was monitored mass spectrometrically. The CO_2 signal,
which is a measure of the CO_2 production rate, is
strongly peaked on both the heating and cooling cycles.
Since the maxima both occur near 573K, we expect to
find maxima in the steady-state CO_2 production rates
near the same temperature. This data was used as a
qualitative guide only. The only significant
differences in the rates are the amplitude, which
is higher during cool-down, and the peak temperature,
which is slightly lower during cool down. These are
ascribed to two differences: (1) the absolute rate
of temperature change and (2) the amounts of adsorbed
CO and oxygen which are present during the transients.

Figure 2 shows data from the variation of P_{O_2}
taken under conditions similar to those of Figure 1.
The origin of the ordinate is defined as the steady
state O_2 pressure measured at 300K. As the sample is
heated the O_2 pressure drops (open circles) goes
through a local minimum and a maximum between 15 and
40s (500 and 800K). The triangles show a predicted
O_2 pressure curve based on CO_2 pressure. Although
one might argue that inaccurate relative sensitivities
can account for some of the difference in these two
curves, it is clear that the local maximum in the
experimental O_2 consumption is not reflected in the
calculated curve based on CO_2 production. As indicated by the difference spectrum (solid line in the
upper portion of Figure 2) there is not a single
scaling factor which will bring any significant
portion of these two curves into coincidence. From
these results we conclude that O penetration into
the subsurface region is an important process,
particularly at high temperatures. The significance
of this process has been confirmed by work in our
own (12) and other laboratories (13).

A central point to be made in connection with
Figures 1 and 2 is that a great deal of qualitative
insight can be gained from relatively simple experiments. The results guide the selection of temperatures for steady-state and transient experiments
which can be analyzed more quantitatively.

Another kind of thermally induced transient is illustrated in Figure 3. This data, taken from work of Charles Campbell and Shei-Kung Shi (14), involves the transient production of CO_2 which occurs when a coadsorbed mixture of CO molecules and O atoms on a Rh surface is heated. The various spectra are for different titration times indicated on the figure (0-20 min); a polycrystalline Rh surface presaturated with oxygen was titrated at 360K for x min. by 9.6 x 10^{-7} Pa (133 Pa = 1 Torr) of CO and, after evacuation, was heated to 750K or higher. With titration time (i.e. prior to flashing) CO_2 is produced. This lowers the amount of chemisorbed oxygen and allows increasing amounts of CO to adsorb. Thus, the areas beneath the curves in Figure 3 reflect qualitatively the mathematical product of the coverage of CO times the coverage of oxygen present at the start of the heating cycle. The former is small at short titration times while the latter is small at long titration times. It is important to note the spectra shown in Figure 3 were generated by heating the surface in vacuum. The area beneath the CO_2 pressure transient varies strongly with titration time indicating that CO_2 is readily produced by reaction of coadsorbed CO and O. In fact, companion CO spectra show that all the chemisorbed CO reacts to make CO_2 so long as there is excess oxygen. By analyzing CO_2, CO and O_2 transients like those of Figure 3, the temperature dependence of the rate of CO_2 production can be evaluated for a relatively wide range of CO and O coverages (θ_{co} and θ_o). This analysis shows that the rate is not describable by a simple LH expression of the form

$$R_{co_2} = A \exp \{- (E_{LH}/RT\} \; \theta_o \; \theta_{co} \qquad (1)$$

Either coverage dependent terms in the rate coefficients or fractional exponents in the coverage terms must be incorporated. Table I summarizes our analysis of several cases using the equation

$$R_{co_2} = A_{LH} \exp \{ - (E_{LH} + \alpha_o \theta_o + \alpha_{co} \theta_{co})/RT\} \; \theta_o^m \; \theta_{co}^n \qquad (2)$$

Although we cannot demonstrate unambiguously which of these, if any, is physically most satisfactory, we prefer model IV because it preserves the exponents of unity on the coverages, as an elementary equation must, because it gives a good fit to the detailed reaction rate versus titration time at 360K (calcula-

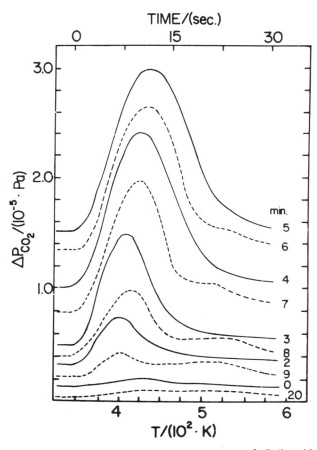

Applications of Surface Science

*Figure 3. CO_2 pressure vs. time during heating a polycrystalline Rh sample cov-
ered with CO and O. These species were deposited at 360 K in varying amounts by
predosing oxygen (15.9 L) and then titrating with CO(g) at 7×10^{-9} torr. The
heating was then done in vacuo. Titration times are marked in minutes. The small
high temperature peak is due to CO interactions with the walls (14).*

Table I

Kinetic analysis.

$$R_{CO_2} = A \exp[-(E_{LH} + \alpha_O \theta_O + \alpha_{CO} \theta_{CO})/RT]\theta_O^m \theta_{CO}^n$$

Case #	m = order in θ_O	n = order in θ_{CO}	$\dfrac{\alpha_O}{\text{kcal mole}^{-1}}$	$\dfrac{\alpha_{CO}}{\text{kcal mole}^{-1}}$
I	1	1	0	0
II	1.6	1	0	0
III	1	0.66	0	0
IV	1	1	-4.4	0
V	1	1	0	2.1

$\ln\dfrac{A}{\text{molec. cm}^{-2}\text{s}^{-1}}$	$\dfrac{E_{LH}}{\text{kcal mole}^{-1}}$	Standard deviation α
41.1 ± 1.02	8.97 ± 0.86	0.299
48.5 ± 0.56	13.31 ± 0.48	0.114[a]
41.0 ± 0.40	8.64 ± 0.34	0.121[a]
47.6 ± 0.40	14.30 ± 0.34	0.073[a]
42.1 ± 0.29	8.60 ± 0.24	0.087[a]

[a] Scaled according to E_{LH} for direct comparison with I.

ted from a plot of θ_o versus titration time), and because it gives the best statistical fit to the data. Perhaps some combination of models IV and V would be even better but our data certainly do not require it. In any case, model V alone does not give a large enough CO_2 production rate when the oxygen coverage is high, so it cannot be used alone.

The interpretation of these coverage-dependent effects involves such ideas as island formation, mixed domains of CO and O, the mobility of CO and O and the adsorption of CO on oxygen-covered regions (1,18,26, 27). A deeper understanding of the roles of these processes comes from isothermal experiments involving pressure transients.

All of the results discussed to this point infer surface coverages from gas phase composition measurements. Although these inferences are soundly based in the case of CO oxidation, we would be much more comfortable with direct measurement of the surface coverage.

Transients Induced by Pressure Changes

The most sophisticated and incisive transient experiments are those derived from modulated molecular beam reactive scattering experiments. Results from such experiments will be discussed below after presentation of data from simpler experiments which not only are consistent with the beam results but also extend the range of temperature-coverage conditions and, in some cases, directly assess the coverage.

Figure 4 shows the results of a very simple experiment involving transient CO_2 production on a polycrystalline Pt foil at 540K (15). The transient was set up by establishing a steady-state reaction with $O_2/CO \approx 10$ and then rapidly closing (≈ 3 sec) the O_2 leak valve into the continuously pumped reaction chamber. In the absence of reaction, the CO leak rate gave a partial pressure of 2×10^{-8} torr. The ensuing transient CO_2 comes from the removal of chemisorbed oxygen atoms present at its origin. From Figure 4 it is immediately clear that the CO_2/CO ratio, and therefore the CO_2 production rate per unit CO pressure, remains nearly constant over the first 2 min of the titration while the oxygen coverage drops off more than a factor of 2. This result clearly eliminates the possibility of reaction through a simple ER process of the form

$$CO(g) + O(a) \rightarrow CO_2(g)$$

for which the rate of CO_2 production would be

$$R_{CO_2} = k_{ER} P_{CO} \theta_o. \tag{3}$$

According to this expression the rate per unit CO
pressure should be proportional to the coverage of
oxygen which it is not. Originally (15) this data
was interpreted in terms of a physisorbed precursor
state largely because no significant accumulation
of chemisorbed CO is possible at this temperature
and it was felt that the CO residence time was
probably very short ($<10^{-5}$ sec). Recent molecular
beam data shows that the residence time for reactive
CO is actually much longer ($\approx 10^{-2} - 10^{-3}$ sec) (16)
meaning that reaction cannot possibly occur in the
physisorbed state (lifetime would be less than 10^{-5}
sec). Theoretical calculations indicate that the
barrier to lateral diffusion of CO across a Pd
surface is quite low, at least along certain
crystallographic directions, (17) so that chemisorbed
CO moves about rapidly and during its surface lifetime will collide with an oxygen atom even when the
oxygen coverage is relatively small.

Before moving on, we should describe the
experimental procedure by which the oxygen coverages
shown in Figure 4 were determined. The total area
beneath the CO_2 versus time curve was taken as a
measure of the initial coverage, θ_{max}, of oxygen
atoms. Integrating this same curve from the beginning
of the transient to time, t, provides a measure of
the coverage removed. By subtraction

$$\theta_o(t) = \theta_{max} - \int_o^t P_{CO_2}(t) \, dt \tag{4}$$

the time dependent coverage is calculated. This
equation assumes that penetration of surface oxygen
into the bulk can be ignored on Pt (18).

Other more direct methods of assessing the
oxygen coverage are available through AES (18), XPS
(19), UPS (20) and other forms of electron spectroscopy. Figure 5 shows a nice example taken from the
work of Tatsuo Matsushima and David Almy (18). In
this work on polycrystalline Pt, a steady-state
reaction was established at various temperatures
with $P_{O_2}/P_{CO} = 10$. At time t = 0, the O_2 pressure

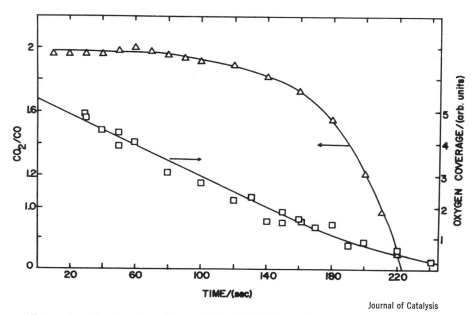

Journal of Catalysis

Figure 4.   The time dependence of CO_2/CO ($\triangle$) and O coverage ($\square$) during the transient titration of chemisorbed oxygen on polycrystalline Pt at 540 K (5).

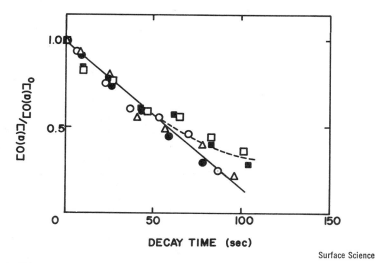

Surface Science

Figure 5.   Decay of reactive oxygen, existing at various temperatures in a steady-state $CO/O_2/Pt$ system, after termination of the O_2 supply. The CO pressure was constant at 1×10^{-8} torr (18). $P_{CO} = 1.3 \pm 0.1 \times 10^{-6}$ Pa. Key: ■, 373 K; □, 413 K; $\triangle$, 473 K; ○, 523 K; and ●, 608 K.

is rapidly reduced to zero and the oxygen Auger signal
is measured as a function of time. The results,
normalized to the initial Auger signal, are shown in
Figure 5 for P_{co} = $1x10^{-8}$torr. Above a relative
coverage of 0.5 the decay is linear with time which
is consistent with a process that is zero-order in
oxygen coverage. At low temperatures and low oxygen
coverages the decay becomes more exponential which is
consistent with a first order dependence on oxygen
coverage. Other transients at a fixed temperature
demonstrate that the rate is first order in P_{co}.
        Figure 6 shows that all of the chemisorbed
oxygen is reactive. Here, a polycrystalline Pt
surface at 370K is exposed to various amounts of O_2
as indicated on the abscissa and, after termination
of the O_2 input, titrated with CO. As indicated
by the four sets of dashed lines in the figure, all
of the oxygen Auger signal vanishes.
        A very important non-transient measurement
related to the carbon monoxide oxidation reaction
is shown in Figure 7. These data show the steady-
state oxygen Auger signal for various CO pressures
and three fixed oxygen pressures. When taken as a
measure of the oxygen coverage these may be compared
to the oxygen coverages determined by transient CO_2
production after terminating the O_2 input in a steady-
state experiment. Although the details cannot be
given here, agreement is very satisfying (18).
        Although the use of Auger spectroscopy to detect
directly oxygen coverages during the course of a
transient or steady-state reaction is very satisfying,
there are limitations. One of these is illustrated
in Figure 7; the bracketed open triangles indicate
the detectable presence of carbon and oxygen on the
surface. This, of course, means that the carbon
monoxide coverage is significant and must be separated
from the oxygen coverage in the analysis. This is
not a simple problem and it is compounded by the fact
that electron beam damage destroys the carbon-oxygen
bond. The solution to this problem lies in the use
of some other form of spectroscopy, like x-ray photoelectron spectroscopy (XPS), which does not destroy
CO and which can separate chemisorbed oxygen and
chemisorbed CO on the basis of chemical shifts.
        Figure 8 gives an example of time dependent XPS
signals observed during the carbon monoxide titration
of chemisorbed oxygen on Ir(111) at 398K. This data
from Henry Weinberg's lab (19) was taken by presatur-
ating a clean Ir surface with oxygen and then titrating
with $2x10^{-7}$torr of CO while following the time

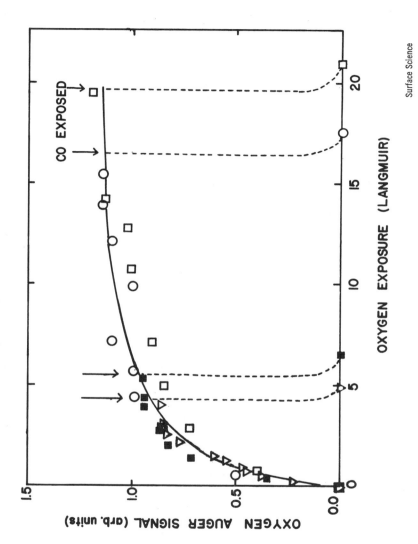

Surface Science

Figure 6. Peak-to-peak oxygen Auger signal (517 eV) as a function of oxygen exposure to poly-crystalline Pt at 370 K. Oxygen pressures (Pa) are 1.13 × 10⁻⁶ (■), 1.29 × 10⁻⁶ (▽), 2.93 × 10⁻⁶ (○), and 6.33 × 10⁻⁶ (□). Vertical arrows indicate points of CO exposure in four separate experiments, and dashed lines connecting to points on the abscissa indicate the loss of the O₂ signal (18).

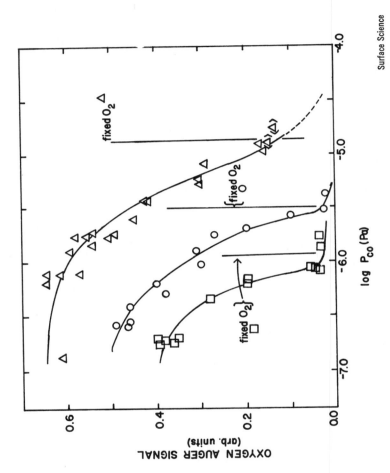

Surface Science

Figure 7. Steady-state oxygen Auger signal on Pt at 483 K as a function of CO pressure for several fixed O pressures. Triangles enclosed in brackets (⟨△⟩) indicate the presence of a detectable concentration of adsorbed carbon monoxide. The fixed oxygen pressures (torr) were 10^{-8} (□), 2.5×10^{-8} (○), and 1.1×10^{-7} (△) (18).

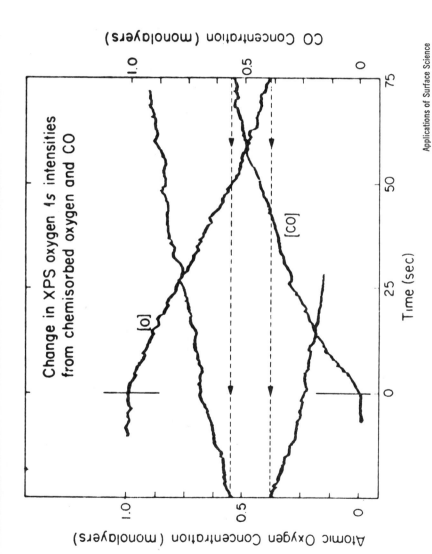

Figure 8. XPS intensities at 529.8 eV, O(a), and 532.0 eV, CO(a), binding energy vs. time for a titration of O(a) by CO(a) on Ir (111) at 398 K. The CO pressure was fixed at 2 × 10⁻⁷ torr (19).

dependence of XPS signals at 529.8, O(a), and 532.0 eV,
CO(a). These results, and others like them, can be
used to compute the time dependent coverages of both
CO and O for a variety of conditions. From the slope
of the oxygen coverage versus time, a CO_2 rate can
also be calculated. In Figure 9 the function Rate/
[O(a)][CO(a)] is plotted as a function of time. The
curvature of this plot shows clearly that a single
Langmuir-Hinshelwood expression cannot be used over
a broad range of coverages. This same kind of
behavior has been reported for Pd (16a) and Rh (14).
The interpretation given for Ir (19) involves two
competing reaction mechanisms -- one involves CO
adsorbed on Ir and the other involves CO adsorbed over
oxygen on Ir.

Recently, Shei-Kung Shi and John Schreifels in
our laboratory (21) have done an interesting transient
experiment involving the titration of adsorbed oxygen
with hydrogen on Ru(001). The results indicate the
caution with which Auger spectroscopy should be used
to follow reactive chemisorbed oxygen concentrations.
On Ru, the H_2 + O(a) reaction is characterized by
long induction times at high oxygen coverages. Figure
10 shows a comparison of the time dependences of
this reaction as followed at 500K using AES and XPS
to measure the surface oxygen. The initial oxygen
coverages were identical. Unfortunately, the H_2O
production rates were so slow that quantitative
measurement of them was not possible. The differences
in Figure 10 are startling, and indicate a rather
strong electron beam effect causing O^+ desorption
and/or activation of the surface oxygen. Although
the details of this are now being worked out, this
example illustrates the care which must be exercised
in interpreting kinetic data using Auger for oxygen
coverages.

Before considering modulated molecular beam
results, we examine a very interesting example of
CO_2 production transients from the work of Madix
and co-workers (22). An example of the data is given
in Figure 11 where the CO_2 mass spectrometer signal
is plotted as a function of time upon exposure to CO
of a Ag(110) surface partially covered with oxygen.
Clearly the slope (proportional to the rate) is
higher in magnitude at the lower temperature. From
an analysis of such curves over a temperature range
from 150 to 400K an apparent activation energy of
-1.1 kcal mole^{-1} is calculated. This is interpreted
as the difference between the activation energy for
reaction and that for desorption of CO, $(E_r - E_d)$.

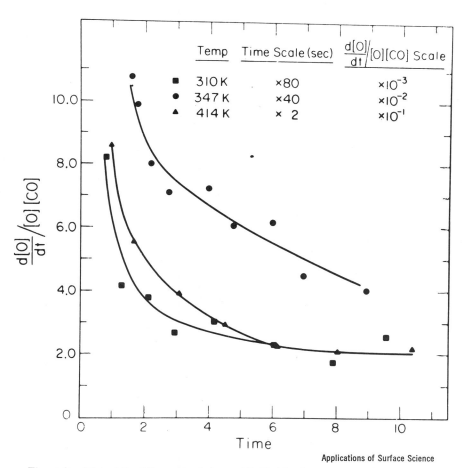

Figure 9.   Plot of the CO_2 production rate (divided by the coverages of oxygen and carbon monoxide) as a function of time for the titration of O by CO on Ir (111).

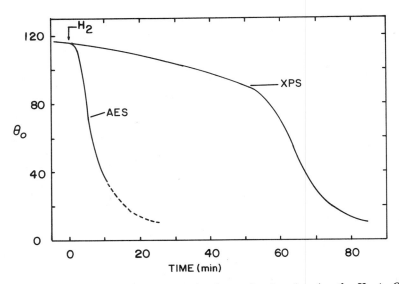

Figure 10.   Oxygen AES and XPS signals vs. titration time for the $H_2 + O(a)$
reaction on Ru (001) at 570 K (21).

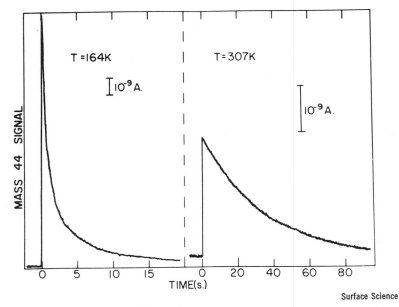

Figure 11. CO_2 signal vs. time for the titration of oxygen chemisorbed on Ag
(110) at 164 K and 307 K (22).

If CO is physisorbed then $E_d \simeq 2$ kcal mole^{-1} and $E_r \simeq$
+1 kcal mole^{-1}. On the other hand if CO is chemisorbed
$E_d \simeq 6$ kcal mole^{-1} and $E_r \simeq 5$kcal mole^{-1}. Since
strongly chemisorbed oxygen atoms and slightly per-
turbed CO molecules are involved, Madix prefers the
latter path. Although this is inferential, it is
very reasonable and certainly is in agreement with
molecular beam work on other transition metals
(described below).
      Throughout this section on pressure transients
we have emphasized electron spectroscopy as a procedure for directly detecting surface species, and,
with difficult calibration, their concentration. It
is important to keep in mind that the detection limit
for these is about 0.01 of a monolayer. Using flash
desorption as a complementary technique this limit
can be extended to 0.001 monolayer in certain cases.
The fact remains that extremely labile chemisorbed
species may be present in kinetically important but
undetectable concentrations. Since residence times
as short as 2×10^{-5} seconds can be determined,
molecular beam techniques, as described below, afford
an alternative but indirect method of measuring the
properties of these very reactive species.
      Just as in gas phase kinetics, reactive molecular
beam-surface scattering is providing important molecular level insight into reaction dynamics. There is
no surface reaction for which such studies have proven
more illuminating than the carbon monoxide oxidation
reaction. For example Len, Wharton and co-workers
(23) found that the product CO_2 exits a 700K Pt
surface with speeds characteristic of temperatures
near 3000K. This indicates that the CO_2 formed by
the reactive encounter of adsorbed species is hurled
off the surface along a quite repulsive potential.
      A very incisive set of experiments on Pd(111)
(16a) and Pt(111) (16b) done in Gerhard Ertl's lab,
show that the Eley-Rideal pathway makes no measurable
contribution to the CO_2 production rate for many low
pressure conditions. In these experiments, a steady-
state CO pressure was established, the O_2 pressure
was modulated and the phase-lag of the modulated CO_2
product signal was measured. The slope of $\ln (\tan\phi)$,
where ϕ is the phase lag, as a function of T^{-1} can
be interpreted in terms of an activation energy
difference $(E_r - E_d)$ between reaction and desorption.
The result for Pt(111) is -10.8 kcal mole^{-1} as shown
in Figure 12 (16b). For an Eley-Rideal pathway in
which a gas phase CO molecule makes a direct or impact
attack on an oxygen adatom, $E_d = 0$ and we require

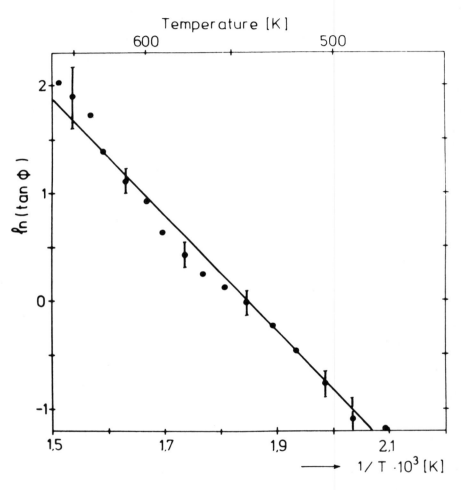

Journal of Chemical Physics

Figure 12.   Logarithmic plot of the tangent of the phase lag as a function of temperature for CO_2 reactively scattered from Pt (111).   The experiment was run with a background CO pressure of 10^{-7} torr and a modulated O_2 beam (10 Hz) which had an equivalent pressure of 6.3×10^{-7} torr.   Key: ●, data; and ——, model, $E_{LH}^* = 24.1$ kcal/mol, $v_4 = 0.11$ cm²/particles s (16b).

E_r < 0 which is unreasonable. If a physisorbed CO is involved, its binding energy would have to exceed 10 kcal mole^{-1} in order for E_r to become positive and it is not reasonable to expect a binding energy of this magnitude. We are thus led to the conclusion that the reaction occurs through an LH pathway.

The same conclusions hold for Pd(111) (16a) where, on the basis of phase shift measurements, $E_r - E_d$ = -9.5 kcal mole^{-1}. In this same very interesting work, transient pressure jump experiments with excellent time resolution were performed. An example is shown in Figure 13 in which a CO beam, with a flux equivalent to P_{CO} = 6x10^{-8} torr impinges on a Pd(111) substrate immersed in a constant ambient P_{O_2} = 1x10^{-7} torr. The important points to note are the induction time at 375K and the transients, with clear maxima noted at 375 and 450K. Since the beam can be turned on in about 1 msec the response shown in Figure 13 is not an artifact of the procedure and we must conclude that more than ER kinetics is involved even under these conditions of high oxygen coverage and relatively low temperatures; an ER path would show no induction time. The maxima observed in Figure 13 are related to the variation of the CO_2 production rate with coverages of CO and O. At short times θ_{co} is low while θ_o is high and the reaction rate grows as CO accumulates. With the passage of time, θ_o declines and θ_{co} continues to grow. The resulting product $\theta_o \theta_{co}$ maximizes when $\theta_o = \theta_{co} = 0.5$. At longer times and lower temperatures θ_{co} inhibits dissociative oxygen adsorption and the rate drops to some steady-state value. At high temperatures, θ_{co} cannot rise to inhibit O_2 adsorption. Consequently, the rate rises upon admission of CO but there is no pronounced maximum.

Finally, we note a very clever isotope experiment done by Tatsuo Matsushima (24). The results are summarized in Figure 14 and show that when ^{13}CO is preadsorbed on Pt and then a mixture of ^{12}CO and O_2 is introduced into the gas phase, the product CO_2 molecules initially formed all are labelled with ^{13}C. In fact the experiment is not quite so simple. An initial $^{13}CO_2/^{12}CO_2$ ratio is measured (unity in Figure 14) which is equal to the $^{13}CO/^{12}CO$ ratio in the preadsorbed CO. The conclusion is that CO_2 is produced by the reaction of adsorbed, rather than gaseous, CO and adsorbed O atoms - i.e. the Langmuir-Hinshelwood process. This conclusion has dominated all of the results presented here and appears to be ubiquitous for all transition metals under all conditions that have been studied to date.

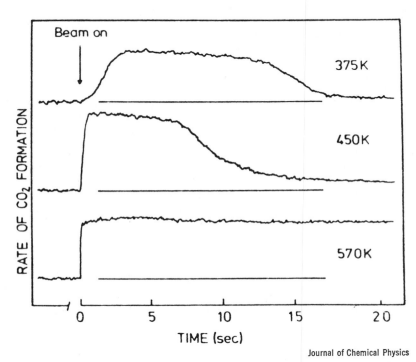

Journal of Chemical Physics

Figure 13.   Transient CO₂ production rates as a function of time for the titration of preadsorbed oxygen on Pd (111) at 375, 450, and 500 K. The experiment was run in a background of 10⁻⁷ torr of O₂, and the CO pressure 6 × 10⁻⁸ torr was introduced at the point indicated (16a).

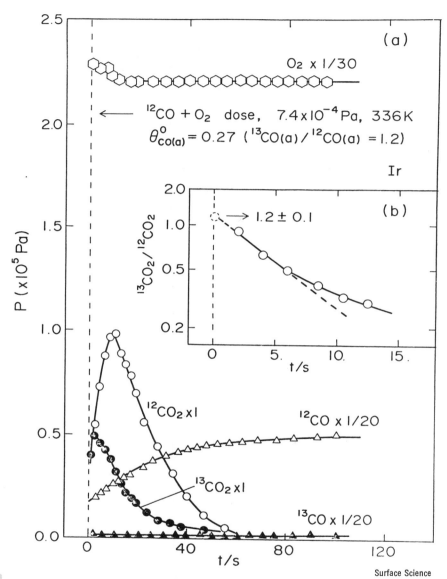

Surface Science

Figure 14.   a: Variation of $^{13}CO_2$ and $^{12}CO_2$ produced by reaction of preadsorbed ^{13}CO and ^{12}CO with a $^{12}CO/O_2$ mixture.   The catalyst was polycrystalline Pt at 300 K, and the ratio of ^{13}CO to ^{12}CO was 1.0.

b: The ratio of $^{13}CO_2$ to $^{12}CO_2$ observed during the initial stages of the reaction (24).

Questions for Future Study

Among the ripest questions for study are those related to the way various modes of motion participate. For example, it is very important to determine the internal energy distribution of the exiting CO_2 molecule and how this varies with the surface structure of both the underlying metal lattice and the chemisorbed layer. The angular dependence of this distribution is also a question ripe for study.

In connection with practical situations where CO oxidation is important, we must also consider the perennial question of how to connect the low pressure results onto those at high pressure. Qualitatively this has been done for the CO oxidation reaction but it would still be worthwhile to attempt a numerical prediction of high pressure results based on low-pressure rate parameters. A very nice paper modeling steady-state CO oxidation data over a supported Pt catalyst at CO and O_2 pressures of several torr has very recently appeared (25). Extension of this work to other systems in warranted and, even though unresolved questions continue to exist, every indication is that the high and low pressure data can be reliably modeled with the same rate parameters if no adsorption – desorption equilibria are assumed.

The role which mobility of surface metal atoms play and the role of oxygen penetration into the lattice is only now beginning to be examined.

Even though widely studied, this reaction continues to defy a full description. Its simplicity in terms of the species involved and the reproducibility with which many of its characteristics have been established make it remain a model system for a wide variety of studies.

Acknowledgement

This work was supported in part by the Robert A. Welch Foundation and the National Science Foundation (CHE 80-05107). A fellowship to JRC awarded by the New Mexico Section of the American Vacuum Society is gratefully acknowledged.

Literature Cited

1. Engel, T; Ertl, G. Advances in Catalysis,
   J. Catalysis 1979, 28, 1.
2. a. Langmuir, I. J. Am. Chem. Soc. 1915, 37, 1162.
   b. Langmuir, I. Trans. Faraday Soc. 1922, 17, 621.
   c. Langmuir, I. Trans. Faraday Soc. 1922, 17, 672.
3. a. Ertl, G; Koch, J. "Proc. 5th Int. Congr. on
      Catalysis"; Hightower, J., Ed.; North Holland:
      Amsterdam, 1973; p. 969.
   b. Hopster, H; Ibach, H.; Comsa, G; J. Catalysis
      1977, 46, 37.
4. Cochran, H. D.; Donnelly, R. G.; Modell, M.;
   Baddour, R. F. "Colloid and Interface Science",
   Vol. III; Kerker, M., Ed.; Academic Press: New
   York, 1976; p. 131.
5. Golchet, A; White, J. M. J. Catalysis 1978, 53,
   266.
6. Sheintuch, M.; Schmitz, R. A. Catal. Rev.--Sci.
   Eng. 1977, 15, 107.
7. Redhead, P. A. Vacuum 1962, 12, 203.
8. Falconer, J. L.; Madix, R. J. J. Catalysis 1977,
   48, 262.
9. Taylor, J. L.; Weinberg, W. H. Surface Sci. 1978,
   78, 259.
10. Gorte, R.; Schmidt, L. D. Surface Sci. 1978, 76,
    559.
11. Close, J. S.; White, J. M. J. Catalysis 1975,
    36, 185.
12. Campbell, C. T.; Foyt, D. C.; White, J. M.
    J. Phys. Chem. 1977, 81, 491.
13. a. Conrad, H.; Ertl, G.; Kuppers, J.; Latta, E. E.
       Surface Sci. 1977, 65, 245.
    b. Ducros, R.; Merrill, R. P. Surface Sci. 1976,
       55, 227.
    c. McCabe, R. W.; Schmidt, L. D. Surface Sci.
       1976, 85, 60.
14. Campbell, C. T.; Shi, S. K.; White, J. M. Appl.
    of Surface Sci. 1979, 2, 382.
15. White, J. M.; Golchet, A. J. Chem. Phys. 1977,
    66, 5744.
16. a. Engel, T.; Ertl, G. J. Chem. Phys. 1978, 69,
       1267.
    b. Campbell, C. T.; Ertl, G.; Kuipers, H.;
       Segner, J. J. Chem. Phys. 1980, 73, 5862.
17. Doyen, G.; Ertl, G. Surface Sci. 1977, 69, 157.
18. Matsushima, T.; Almy, D. B.; White, J. M. Surface
    Sci. 1977, 67, 89.

19.   Zhdan, P. A.; Boreskov, G. K.; Boronin, A. I.;
      Schepelin, A. P.; Withrow, S. P.; Weinberg, W. H.
      Appl. of Surface Sci. 1979, 3, 145.
20.   Matsushima, T.; Hashimoto, M.; Toyoshima, I.
      J. Catalysis 1979, 58, 303.
21.   Shi, S.-K.; Schreifels, J.; White, J. M., in
      preparation.
22.   Bowker, M.; Barteau, M. A.; Madix, R. J.   Surface
      Sci. 1980, 92, 528.
23.   Becker, C. A.; Cowin, J. P.; Wharton, L.;
      Augerbach, D. J.   J. Chem. Phys. 1977, 67, 3394.
24.   a. Matsushima, T.   J. Catalysis 1978, 55, 337.
      b. Matsushima, T.   Surface Sci. 1979, 87, 665.
25.   Herz, R. K.; Marin, S. P.   J. Catalysis 1980, 65,
      281.
26.   Conrad, H.; Ertl, G.; Kuppers, J.   Surface Sci.
      1981, 76, 323.
27.   Barteau, M. A.; Ko, E. I.; Madix, R. J.   Surface
      Sci. 1981, 104, 161.

RECEIVED June 26, 1981.

The Dynamic Behavior of Three-Way Automotive Catalysts

RICHARD K. HERZ

General Motors Research Laboratories, Physical Chemistry Department, Warren, MI 48090

The composition of engine exhaust entering a three-way catalytic converter oscillates about the stoichiometric point at frequencies of 0.5 to 2 Hz when the engine is operating under feedback control. Time-averaged conversion measurements have shown that three-way catalysts do not respond instantaneously under these rapidly varying conditions. Recently, time-resolved conversion measurements have permitted direct observation of the dynamic responses of three-way catalysts. After a step change from lean to rich conditions, for example, CO conversions over a catalyst may take several seconds to decay to a new, lower steady-state level. One possible explanation of this response is reaction of the CO in rich exhaust with oxygen that was stored on the catalyst during the preceding exposure to lean exhaust. Other measurements have demonstrated that the oxygen content of a base metal-containing three-way catalyst can vary with exhaust composition at rates which are sufficient to explain the observed dynamic responses of CO conversion.

Three-way catalysts are used in most 1981 gasoline-fueled automobiles to lower the levels of NO, CO, and hydrocarbons in engine exhaust. These catalysts normally operate under dynamic conditions: catalyst temperature increases rapidly after the engine starts (during catalyst "warmup"), and exhaust flowrate and composition fluctuate under most modes of operation. The warmup of automotive catalysts is reasonably well understood (1). The operation of three-way catalysts in the dynamic exhaust environment after warmup is complex and less well understood.

In this paper, we summarize the progress that has been made toward understanding the behavior of three-way catalysts in exhaust of rapidly varying composition. After a brief description of the conditions under which three-way catalysts operate

0097-6156/82/0178-0059$05.00/0

in an automobile, we discuss time-averaged NO and CO conversions
measured under laboratory conditions which simulate actual
operation. Next, we discuss time-resolved CO conversions measured during step-response experiments. Some of the step-response
results may be explained by changes in the amount of oxygen
bound to the catalyst, and measurements of these changes are
reviewed. We conclude by suggesting approaches to be taken in
future studies.

Conditions During Automobile Operation

Three-way catalysts are able to reduce NO as well as oxidize CO and hydrocarbons when the exhaust composition is held
near the stoichiometrically balanced composition, or "stoichiometric point." This control of exhaust composition is accomplished, after the initial warmup period, through the use of
the feedback control system illustrated in Figure 1 (2, 3, 4).
An oxygen sensor signals whether the exhaust entering the
converter is "rich" (net reducing) or "lean" (net oxidizing).
The voltage response of an oxygen sensor is shown in Figure 2.
A microcomputer periodically reads the sensor signal and adjusts
the fuel control signal to bring the air-fuel ratio, and thus
the exhaust composition, closer to the stoichiometric point.
The exhaust composition cannot be held exactly at the stoichiometric point, however. As a result of the switch-like response
of the oxygen sensor and the time required for flow from the
fuel metering point through the engine to the sensor, the control system exhibits limit-cycle behavior. The exhaust composition oscillates about the stoichiometric point at a frequency
which increases with exhaust flowrate over the range 0.5 to 2
Hz. This oscillation, or cycling, is illustrated by the oxygen
sensor signal shown in Figure 3. In addition to these fairly
regular oscillations in exhaust composition, transients in
composition occur during rapid vehicle acceleration and de-
celeration. Depending on the response of the catalyst, lean
excursions from the stoichiometric point may result in increased
NO emissions and rich excursions may result in increased CO and
hydrocarbon emissions.

Time-Averaged Conversion Measurements

The performance of a three-way catalyst under dynamic
conditions is often studied by measuring the time-averaged
reactant conversions that are obtained as the feedstream composition is deliberately cycled. The variables in these cycled
studies include time-averaged feedstream composition, cycling
frequency, cycling amplitude, and catalyst composition. Cycled
studies are performed in engine-dynamometer laboratories by
switching the air-fuel ratio between two settings at selected
frequencies (5, 6, 7). Cycled studies also have been performed

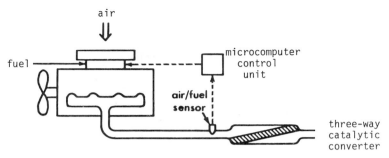

Figure 1.   Schematic diagram of an engine and emission control system. The microcomputer also reads signals from sensors measuring other engine operating parameters. In some emission control systems, the three-way catalyst is followed by supplementary air injection and an oxidizing catalyst to provide additional control of CO and hydrocarbon emissions.

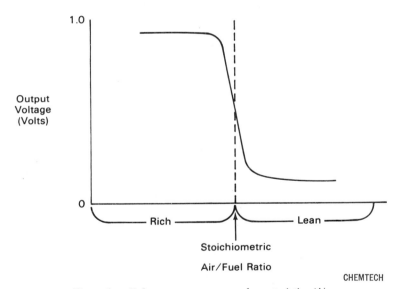

Figure 2.   Exhaust oxygen sensor characteristics (4).

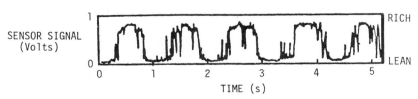

Figure 3.   Oxygen sensor signal showing oscillation of the exhaust composition around the stoichiometric point during feedback control of the engine.

(6) using the laboratory reactor system developed by Schlatter, Sinkevitch, and Mitchell (8), shown in Figure 4. Two simulated exhaust streams of different compositions are alternately fed to the reactor in this system. Time-averaged inlet and outlet compositions are measured in these cycled studies because of the limited response rates of conventional exhaust analyzers. Figure 5 presents CO and NO conversion data which are characteristic of the cycled performance of three-way catalysts. Note that cycling results in lower average conversions for mean air-fuel ratios near the stoichiometric point. The extent to which cycling affects average conversions over a catalyst has been found to be sensitive to changes in the composition of the catalyst and to aging of the catalyst in engine exhaust (9). Since cycling often results in the lowering of conversions, there is a substantial incentive to study the dynamic behavior of three-way catalysts. One incentive is that an understanding of the effects of catalyst composition on dynamic behavior may lead to the formulation of catalysts which give required conversions under cycled air-fuel ratio conditions with the use of minimum amounts of noble metals.

Note in Figure 5 that average conversions are dependent on cycling frequency. Some of this effect is related to mixing in the feedstream which decreased the amplitude of the inlet composition oscillations as the frequency was increased. However, this frequency effect does indicate that the dynamic response of the catalyst was complex such that conversions over the catalyst did not change instantaneously as the feedstream composition changed. This complexity is shown more specifically for CO by Figure 6. The solid line gives conversions measured under steady-state conditions. The dotted line represents the average conversions that would be expected if the outlet CO concentration instantaneously reached a new steady-state value as the inlet composition is cycled as a square wave about the mean air-fuel ratio. Actual conversions obtained when cycling at a frequency of 0.25 Hz are given by the dashed line. The result that the measured cycled conversions are higher than the predicted conversions implies that conversions during the rich half of a cycle are higher than would be obtained under these rich conditions at steady-state.

Dynamic Gas-Phase Measurements

Although time-averaged conversion measurements provide valuable information, a detailed analysis of the dynamic behavior of three-way catalysts requires direct observation of catalyst responses to rapid changes in exhaust composition. Some transients over three-way catalysts in simulated exhaust occur slowly enough that they can be observed with conventional exhaust analyzers. Schlatter and Mitchell (9) have studied the step responses of $Pt/Rh/Ce/Al_2O_3$ catalysts in simulated exhaust.

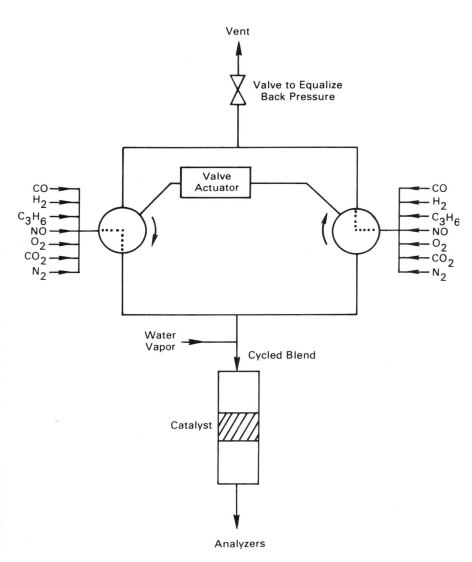

Figure 4. Laboratory reactor system used to measure the performance of catalysts under cycled conditions. Two simulated exhaust streams of differing compositions are alternately fed to the reactor (8).

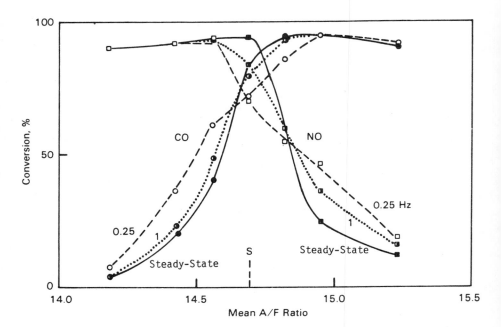

Figure 5. Time-averaged CO and NO conversions measured using the laboratory reactor system shown in Figure 4. A fresh, pelleted $Pt/Rh/Al_2O_3$ catalyst was operated at a middle-bed temperature of 820 K and a space velocity of 52,000 h^{-1} (STP). The feedstreams simulated exhaust that would be obtained with various engine air–fuel ratios (A/F) but did not contain SO_2. The feedstream compositions were cycled at 0.25 and 1 Hz at an amplitude of ± 0.25 A/F about the mean A/F. For the curves labeled Steady-State, conversions were measured with feedstreams at the mean A/F values (8).

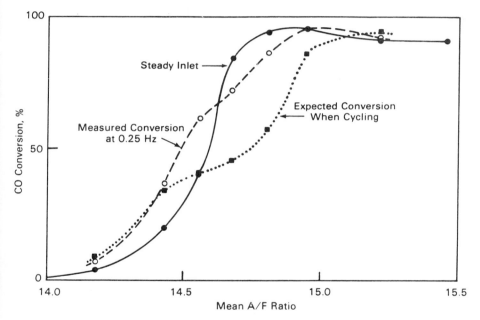

Figure 6.   CO conversions measured under steady and cycled (± 0.25 A/F) conditions. Expected conversions were calculated by assuming that the catalyst would respond instantaneously to changes in feedstream composition (8).

They found that CO conversions can take up to 15 minutes to decrease to a new steady-state level after a change from lean to
rich conditions at a space velocity of 100,000 h^{-1} (STP) and a
middle-bed temperature of 823 K. The transient enhancement in
CO conversion was found to be due to the water-gas shift reaction catalyzed by Rh. An oxidized form of Rh was proposed to be
active for the water gas shift reaction, with reduction of the
oxidized Rh occurring slowly in the rich feedstream. Ce pro-
longed the transient enhancement in CO conversion, apparently by
retarding the rate of Rh reduction. When SO_2 was added to the
feedstream, CO conversions reached a new steady-state level in
less than 10 s, or faster than the response rate of the analysis
system. SO_2 is present in exhaust at a level of about 20 ppm
and poisons the water-gas shift reaction on the Group VIII transition metals (10). Cooper and Keck (11) have studied the step
responses of Ni-containing three-way catalysts in SO_2-free simulated exhaust. At a space velocity of 50,000 h^{-1} (STP) and a
catalyst inlet temperature of 813 K, they observed transients in
conversion over periods up to 10 minutes long. The water-gas
shift activity of the Ni and changes in the amount of oxygen
bound to the Ni were proposed to affect the observed step responses of the catalysts.

Conversions over three-way catalysts change more rapidly
after a change in inlet composition when the feedstream is
actual engine exhaust rather than SO_2-free simulated exhaust.
The difference between the catalyst responses in the two feedstreams may be related to factors which are yet unknown in
addition to SO_2 poisoning of the water-gas shift reaction.
Instruments which have 10%-to-90% response times of less than
0.1 s are required to measure the dynamic response of a catalyst
in engine exhaust.

As we have seen in Figure 3, the oxygen sensor used in the
engine control system is fast enough to respond to most of the
important changes in exhaust composition. However, the oxygen
sensor cannot be used to measure concentrations of reactants in
exhaust since the sensor's output is a nonlinear function of the
ratio of oxidizing species to reducing species. Meitzler (12)
has used oxygen sensors placed before and after a catalytic
converter to study the dynamic response of a three-way catalyst.
The catalyst was found to retain oxidizing species into the
initial periods of rich portions of air-fuel ratio cycles.

Infrared diode laser spectroscopy has been used for the
measurement of hydrocarbon and CO concentrations in exhaust (13,
14, 15). The adsorption path length, and thus the absorption
cell volume, required for hydrocarbon measurement is rather
large, limiting the time-resolution of the measurement. The
absorption path length required for CO measurement, however, is
relatively short and approximately equal to the diameter of a
standard exhaust pipe. This allows CO to be measured with high
time-resolution by an infrared laser beam passed through an

exhaust pipe. The path length required for NO measurement may
be short enough to allow dynamic measurement of this species.
O_2 cannot be measured by conventional infrared spectroscopy.
     Figure 7 is a schematic diagram of the apparatus used for
dynamic CO measurements. An infrared diode laser beam is split
and components pass through the exhaust at the inlet and outlet
of a standard catalytic converter. The laser beam is chopped at
a high frequency and lock-in amplifiers are used to separate the
detector signal due to the laser from that due to the high
infrared background near the engine. A minicomputer is used to
record the carburetor or fuel-injector control signal, the two
laser signals, and signals from oxygen sensors placed at various
points in the exhaust flow system. The engine can be operated
under normal feedback control, or the minicomputer can be used
to control the air-fuel ratio. The 10%-to-90% response-time of
this system is 0.025 s, which is sufficiently fast to follow
accurately any CO concentration change which occurs in automotive exhaust.
     Figure 8 shows the CO concentrations measured at the inlet
and outlet of the catalytic converter during feedback control of
the exhaust composition. The flow time between the inlet and
outlet measurement points was 0.06 s at the space velocity of
50,000 h^{-1} (STP) and converter temperature of 773-823 K. The
converter contained a pelleted $Pt/Pd/Rh/Ce/Al_2O_3$ catalyst which
had been aged for 30,000 km on an automobile. The relatively
large, low frequency (ca. 1 Hz) oscillation in the inlet CO
level was due to limit-cycling of the control system. The lower
amplitude, higher frequency (ca. 10 Hz) oscillations resulted
from two superimposed components: (1) a component due to modu-
lation of the fuel-metering rod in the carburetor, and (2) a
component due to maldistribution of fuel to the engine's cylinders (15). The exhaust composition fluctuations shown by the CO
measurements are also present in the oxygen sensor signal in
Figure 3. The high frequency CO oscillations were present at
the converter outlet during the rich periods of the limit-
cycles; little or no CO broke through the converter when the
exhaust was lean.
     These results are direct measurements of the CO concentra-
tions which are present at the inlet to a three-way converter
under realistic operating conditions. The concentrations of
other exhaust components at the converter inlet can be estimated
using concentration crossplots, for each component versus CO,
obtained from time-averaged measurements at constant air-fuel
ratio settings.
     One of the goals of an analysis of three-way dynamic behav-
ior is to predict outlet concentrations given rapidly varying
inlet concentrations, such as those shown in Figure 8. The
analysis is made easier, however, when the dynamic conditions
are simplified to step changes and single frequency, constant
amplitude oscillations in air-fuel ratio. Figure 9A shows CO

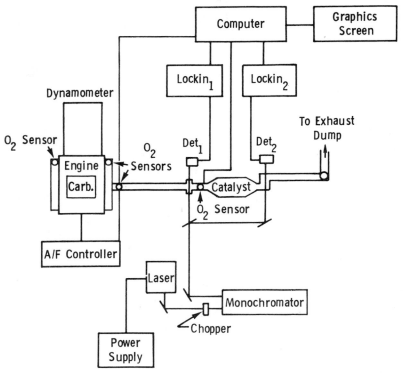

Society of Automotive Engineers

Figure 7.   Schematic diagram of the apparatus used to study the dynamic behavior of three-way catalysts in engine exhaust (14, 15). This apparatus was used to make the measurements shown in Figures 8 and 9.

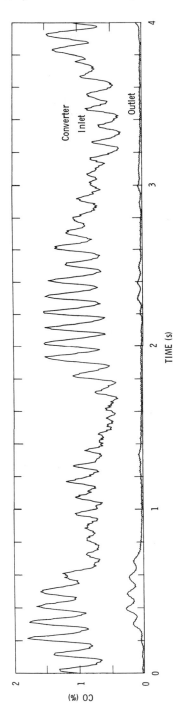

Society of Automotive Engineers

Figure 8. CO concentrations measured at the inlet and outlet of a three-way catalytic converter during feedback control of the exhaust composition (15).

concentrations measured as the air-fuel ratio control signal was
stepped from rich to lean over a pelleted Pt/Pd/Rh/Ce/Al$_2$O$_3$
catalyst at a space velocity of 50,000 h^{-1} (STP). Before use,
the catalyst was aged for 100 h on an engine-dynamometer to
simulate 6400 km of exposure to automobile exhaust. Although
difficult to see in this figure, the CO at the converter outlet
actually dropped faster than at the inlet (after considering the
flow time between the measurement points). This is because the
inlet O$_2$ concentration (not measured) increased as the inlet CO
concentration dropped so that the conversion of CO increased at
the same time that CO eluted. This rich-to-lean CO response was
found to be insensitive to catalyst composition.

A much more interesting response is shown in Figure 9B.
Here, the air-fuel ratio setting was stepped from lean to rich.
The inlet CO trace is roughly a mirror image of that in Figure
9A. The outlet CO level, however, took a much longer time to
reach a new steady-state level than for the rich-to-lean step.
At the maximum time shown in Figure 9B, the outlet CO level had
only risen to about 60% of the rich steady-state outlet level,
which can be seen on the left side of Figure 9A. Approximately
25 s were required for the outlet CO to reach the new steady-
state level after the lean-to-rich step. This time is much
shorter than that mentioned above for catalysts in SO$_2$-free
simulated exhaust, but is still long with respect to the periods
of the exhaust composition oscillations observed during actual
automotive operation.

The dashed curve in Figure 9B gives the approximate, smoothed result that would have been obtained if the catalyst response
were instantaneous. The area between the dashed curve and the
actual response represents additional CO conversion due to the
noninstantaneous dynamic response of the catalyst. This type of
response is desirable because it will lead to low CO emissions
when the air-fuel ratio cycles about the stoichiometric point.

The additional CO conversion that is observed after a
change from lean to rich exhaust may be due to a temporarily
enhanced water-gas shift activity as proposed by Schlatter and
Mitchell (9), and may also be due to reaction of CO with oxygen
bound to Ce in the catalyst.

On the basis of time-averaged conversion measurements and
laboratory measurements using single component gases, Shelef and
coworkers (5, 16) proposed that the "storage" of oxygen by base
metals is an important factor in the dynamic performance of
three-way catalysts. If this mechanism is important for a
catalyst, then the oxygen content of the catalyst should change
by measurable amounts after sudden changes in exhaust composition.

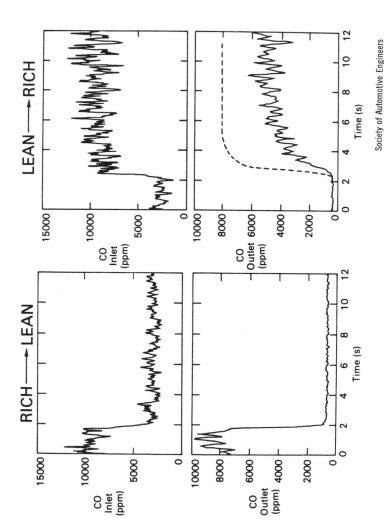

Society of Automotive Engineers

Figure 9.   A: CO concentrations measured over a pelleted Pt/Pd/Rh/Ce/Al$_2$O$_3$ catalyst at a space velocity of 50,000 h^{-1} (STP) as the engine A/F control setting was stepped from rich (A/F = 14.1) to lean (A/F = 15.5).

B: CO concentrations measured as the A/F control setting was stepped from lean (A/F = 15.5) to rich (A/F = 14.1) (14). The dashed curve gives the approximate, smoothed result that would have been obtained if the catalyst response were instantaneous.

Measurement of Catalyst Oxygen Content

Kaneko, et al. (17) have made measurements of the oxygen content of a Pt/Rh/Al$_2$O$_3$ catalyst after various exposures to exhaust. After reduction of the catalyst with CO, the oxygen content of the catalyst was found to increase with exposure to lean exhaust. The data obtained were used to develop a procedure for calculating emissions under cycled air-fuel ratio conditions. The calculations could explain average conversion measurements that could not be explained by assuming the catalyst responded instantaneously to air-fuel ratio changes.

The catalyst studied by Kaneko and coworkers contained only Pt and Rh. Most three-way catalysts, however, also contain base metals in larger amounts than the precious metals (5, 6, 7, 9, 11). We have measured changes in the oxygen content of a Ce-containing three-way catalyst after various exposures to exhaust (18).

The apparatus that was used in our study is shown in Figure 10. After the desired exposure of the 5 g catalyst bed to exhaust, the reactor was flushed with N$_2$ for 10 min, and then the oxygen content of the catalyst bed was determined in the following procedure. First, both solenoid valves at the inlet of the reactor were closed and a small flow of N$_2$ was injected into the reactor through a 4-port switching valve. The switching valve was then turned to inject a stream of 2.3% O$_2$/0.1% Ar/He into the reactor as the flow from the reactor was analyzed continuously with the mass spectrometer. Ar was monitored to determine the elution curve of an inert species, and it also served as an internal concentration standard in the CO$_2$ measurement. The elution curve of O$_2$ relative to that of Ar was used to determine the amount of O$_2$ that reacted with the catalyst. A small amount of O$_2$ reacted with carbon on the catalyst to form CO$_2$, which was also measured (H$_2$O formation was negligible). The oxygen content of the catalyst after exposure to exhaust was calculated by subtracting the net oxygen uptake (measured uptake less the amount reacted to form CO$_2$) from the maximum oxygen content, or oxygen capacity of the catalyst bed. The oxygen capacity of the catalyst bed is defined as the maximum amount of oxygen that was retained by the catalyst after treatment with O$_2$ and that could be removed by reaction with CO.

The rate of increase of oxygen content after a rich-to-lean step change in air-fuel ratio was studied using the following procedure. The reactor flow was switched to N$_2$ after stabilization of the catalyst in exhaust at an air-fuel ratio of 14.1, a catalyst inlet temperature of 680 K, and a space velocity of 110,000 h^{-1} (STP). Next, the air-fuel ratio was changed to a setting of 15.1 while the exhaust bypassed the reactor. Then the reactor was given a pulse of lean exhaust by switching from N$_2$ to exhaust for a specified period, and then back to N$_2$.

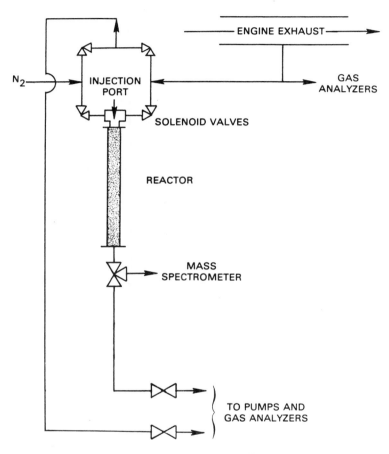

I&EC Product Research and Development

Figure 10. Schematic diagram of the apparatus used to measure changes in the oxygen content of a catalyst with changes in exhaust air–fuel ratio. The 2.54-cm i.d. tubular reactor contained 5 g of catalyst and was heated by an electric furnace (18).

After this exposure, the oxygen content of the catalyst was
measured.

Figure 11 shows the results of these experiments for the
same pelleted $Pt/Pd/Rh/Ce/Al_2O_3$ catalyst used for the data in
Figure 9. The oxygen content of the catalyst bed increased
linearly with pulse duration and reached the steady-state lean
level within 0.5 s. The oxygen capacity of the catalyst was
associated primarily with oxidation of the 190 µmol of Ce contained in each gram of catalyst; the catalyst contained only
8 µmol of precious metal per gram. A comparison of the oxygen
capacity to the amount of Ce in the catalyst suggests that about
76% of the Ce could change between the +3 and +4 oxidation
states (the most common oxidation states of Ce). The identities
of the Ce compounds which undergo oxidation and reduction in
exhaust are not known, however, dispersed hydroxides and oxy-
hydroxides are likely candidates (18).

The rate of decrease of oxygen content of the catalyst
after a lean-to-rich step change was studied using the procedure
described above with the lean and rich exposures reversed.
Figure 12 presents the results of these experiments. After
exposure to rich exhaust for 1.0 s, the oxygen content of the
catalyst bed decreased 36% of the way from its lean steady-state
level to its rich steady-state level. This change in oxygen
content could correspond to a conversion of 58% of the CO and H_2
in the 1.0 s pulse of rich exhaust (hydrocarbons are less reactive with the oxygen held by the catalyst than CO and H_2). The
difference in catalyst oxygen content between the rich and lean
steady-states is sufficiently large to account for the extra
conversion obtained in the CO step response experiment shown in
Figure 9B. In addition, the rates of change of catalyst oxygen
content shown in Figures 11 and 12 are sufficiently fast to
affect the performance of the catalyst when the air-fuel ratio
is cycled about the stoichiometric point at frequencies on the
order of 1 Hz.

Future Studies

The studies described in the preceding two sections have
identified several processes that affect the dynamic behavior of
three-way catalysts. Further studies are required to identify
all of the chemical and physical processes that influence the
behavior of these catalysts under cycled air-fuel ratio conditions. The approaches used in future studies should include (1)
direct measurement of dynamic responses, (2) mathematical
analysis of experimental data, and (3) formulation and valida-
tion of mathematical models of dynamic converter operation.
Such models must be used in conjunction with a model of converter warmup (1) and a model of catalyst deactivation (6) when
designing the distribution of active components in a catalyst
and converter and the engine control strategy.

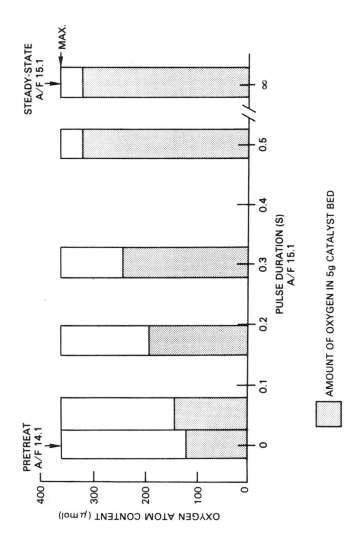

I&EC Product Research and Development

Figure 11. Rate of increase in the oxygen content of a pelleted Pt/Pd/Rh/Ce/Al₂O₃ catalyst in lean exhaust at a space velocity of 110,000 h⁻¹ (STP) and a catalyst inlet temperature of 680 K (18).

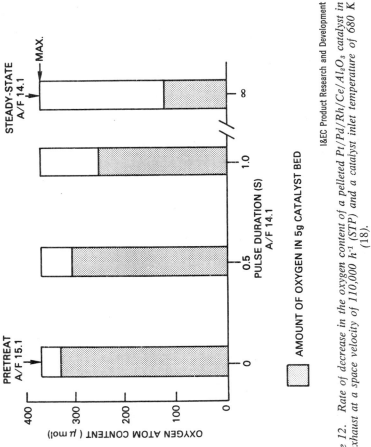

Figure 12. Rate of decrease in the oxygen content of a pelleted Pt/Pd/Rh/Ce/Al$_2$O$_3$ catalyst in rich exhaust at a space velocity of 110,000 h^{-1} (STP) and a catalyst inlet temperature of 680 K (18).

I&EC Product Research and Development

Literature Cited

1. Oh, S. H.; Cavendish, J. C.; Hegedus, L. L. "Mathematical Modeling of Catalytic Converter Lightoff: Single Pellet Studies"; AIChE J., 1980, 26, 935.

2. Canale, R. P.; Winegarden, S. R.; Carlson, C. R.; Miles, D. L., SAE Paper No. 780205, SAE Trans. 1978, 87, 843.

3. Seiter, R. E.; Clark, R. J., SAE Paper No. 780203, SAE Trans. 1978, 87, 828.

4. Hegedus, L. L.; Gumbleton, J. J. CHEMTECH 1980, 10, 630.

5. Gandhi, H. S.; Piken, A. G.; Shelef, M.; Delosh, R. G., SAE Paper No. 760201, SAE Trans. 1976, 85, 901.

6. Hegedus, L. L.; Summers, J. C.; Schlatter, J. C.; Baron, K. J. Catal. 1979, 56, 321.

7. Falk, C. B.; Mooney, J. J. "Three-Way Conversion Catalysts: Effect of Closed-Loop Feed-Back Control and Other Parameters on Catalyst Efficiency"; SAE Paper No. 800462, 1980.

8. Schlatter, J. C.; Sinkevitch, R. M.; Mitchell, P. J. "A Laboratory Reactor System for Three-Way Catalyst Evaluation"; GM Research Publication GMR-2911; presented at 6th N. Amer. Mtg. Catal. Soc., Chicago, IL, March 1979.

9. Schlatter, J. C.; Mitchell, P. J. Ind. Eng. Chem. Prod. Res. Dev. 1980, 19, 288.

10. Joy, G. C.; Molinaro, F. S.; Lester, G. R. "Water-Gas Shift and Steam-Reforming Ability of Group VIII Metals in Simulated Automotive Exhaust"; UOP Inc. report presented at 6th N. Am. Mtg. Catal. Soc., Chicago, IL, March 1979.

11. Cooper, B. J.; Keck, L. "NiO Incorporation in Three Way Catalyst Systems"; SAE Paper No. 800461, 1980.

12. Meitzler, A. H. "Application of Exhaust-Gas-Oxygen Sensors to the Study of Storage Effects in Automotive Three-Way Catalysts"; SAE Paper No. 800019, 1980.

13. Hill, J. C.; Majkowski, R. F. "Time-Resolved Measurement of Vehicle Sulfate and Methane Emissions with Tunable Diode Lasers"; SAE Paper No. 800510, 1980.

14. Sell, J. A.; Herz, R. K.; Monroe, D. R. "Dynamic Measure-
     ment of Carbon Monoxide Concentrations in Automotive Exhaust
     Using Infrared Diode Laser Spectroscopy"; SAE Paper No.
     800463, 1980.

15. Sell, J. A.; Herz, R. K.; Perry, E. C. "Time-Resolved
     Measurement of Carbon Monoxide in the Exhaust of a Computer
     Command Controlled Engine"; SAE Paper No. 810276, 1981.

16. Yao, H. C.; Shelef, M. J. Catal. 1976, 44, 392.

17. Kaneko, Y.; Kobayashi, H.; Komagome, R.; Hirako, O.;
     Nakayama, O., SAE Paper No. 780607, SAE Trans. 1978, 87,
     2225.

18. Herz, R. K. "The Dynamic Behavior of Automotive Catalysts.
     I. Catalyst Oxidation and Reduction"; Ind. Eng. Chem.
     Prod. Res. Dev. 1981, in press.

RECEIVED July 28, 1981.

Dynamics of High-Temperature Carbon Monoxide Chemisorption on Platinum–Alumina by Fast-Response IR Spectroscopy

S. H. OH and L. L. HEGEDUS[1]

General Motors Research Laboratories, Warren, MI 48090

Since the early work of Langmuir ($\underline{1}$), the chemisorption of carbon monoxide on platinum surfaces has been the subject of numerous investigations. Besides its scientific interest, an understanding of CO chemisorption on Pt is of considerable practical importance; for example, the catalytic reaction of CO over noble metals (such as Pt) is an essential part of automobile emission control.

There is a wealth of information available on CO chemisorption over single-crystal and polycrystalline platinum surfaces under ultrahigh-vacuum conditions; research efforts in this area have gained a significant momentum with the advent of various surface analysis techniques (e.g., $\underline{2}$–$\underline{8}$). In contrast, CO chemisorption on supported platinum catalysts (e.g., $\underline{9}$, $\underline{10}$, $\underline{11}$) is less well understood, due primarily to the inapplicability of most surface-sensitive techniques and to the difficulties involved in characterizing supported metal surfaces. In particular, the effects of transport resistances on the rates of adsorption and desorption over supported catalysts have rarely been studied.

Transmission infrared spectroscopy is one of the few techniques applicable to the in situ study of supported catalyst systems (e.g., $\underline{12}$). Much of the evidence concerning the nature of the adsorbed states of CO over supported Pt catalysts has been obtained from infrared spectroscopic studies. For example, it has been shown ($\underline{13}$, $\underline{14}$) that CO can chemisorb on supported Pt catalysts either in a linear or in a bridged form. At high temperatures, however, the absorption band due to the bridge-bonded CO disappears (e.g., $\underline{15}$), and thus the linear form of adsorbed CO represents the dominant state, giving rise to a strong, well-defined infrared band at a frequency of about 2070 cm^{-1}. This provides a convenient means of monitoring the surface concentration of CO under realistic operating conditions by observing the intensity of the associated infrared band.

[1] Current address: W. R. Grace & Company, Washington Research Center, Columbia, MD 21044.

0097-6156/82/0178-0079$06.25/0

Recently there has been a growing emphasis on the use of transient methods to study the mechanism and kinetics of catalytic reactions (16, 17, 18). These transient studies gained new impetus with the introduction of computer-controlled catalytic converters for automobile emission control (19); in this large-scale catalytic process the composition of the feedstream is oscillated as a result of a feedback control scheme, and the frequency response characteristics of the catalyst appear to play an important role (20). Preliminary studies (e.g., 15) indicate that the transient response of these catalysts is dominated by the relaxation of surface events, and thus it is necessary to use fast-response, surface-sensitive techniques in order to understand the catalyst's behavior under transient conditions.

In this paper we will first describe a fast-response infrared reactor system which is capable of operating at high temperatures and pressures. We will discuss the reactor cell, the feed system which allows concentration step changes or cycling, and the modifications necessary for converting a commercial infrared spectrophotometer to a high-speed instrument. This modified infrared spectroscopic reactor system was then used to study the dynamics of CO adsorption and desorption over a Pt-alumina catalyst at 723 K (450°C). The measured step responses were analyzed using a transient model which accounts for the kinetics of CO adsorption and desorption, extra- and intrapellet diffusion resistances, surface accumulation of CO, and the dynamics of the infrared cell. Finally, we will briefly discuss some of the transient response (i.e., step and cycled) characteristics of the catalyst under reaction conditions (i.e., $CO + O_2$).

Fast-Response Infrared Reactor System

a.   Reactor cell. The objective was to construct a rugged infrared cell which can be operated at high temperatures (about 500°C), which has a small enough volume to allow fast response characteristics, and which behaves as a well-mixed reactor (CSTR). The reactor body was manufactured of a pair of Varian nonrotatable blank Conflat flanges, according to a suggestion of Bell (21). These flanges were machined such that a small cavity was created for placement of a thin catalyst disc. A thin thermocouple was placed in direct contact with the disc.

The disc holder was machined to fit into the stainless steel flange in such a way that it directs the gas to consecutively sweep both faces of the catalyst disc, with an expansion volume in-between. This configuration provided good gas-phase mixing in the cell, thus allowing the reactor to be characterized as a CSTR. This mode of internal mixing eliminates the need for internal moving parts or external recycle loops and pumps.

The reactor assembly was heated by electric heaters. The maximum operating temperature is determined by the window construction. Sapphire windows (from EIMAC), brazed into Kovar sleeves, were used; the sleeves were then welded directly into the stainless steel reactor housing. We found that the cell so constructed was capable of trouble-free, continuous operation at 450°C; operations at somewhat higher temperatures are probably still possible but were not explored. Sapphire was chosen as a window material because it is insensitive to water vapor and is transparent in the wave number range of our interest (about 2400 cm^{-1} to 2000 cm^{-1} in these experiments). Moreover, the thermal expansion characteristics of the reactor were found to match well with those of the window fixture.

The catalyst was prepared by impregnating γ-alumina (Alon) to incipient wetness using an aqueous solution of $H_2(PtCl_6)$. After impregnation, the powder was dried, and calcined in air at 773 K (500°C) for 2 h. The infrared disc was prepared by compressing 0.08 g of the catalyst powder at 58 840 N. The properties of the catalyst disc are listed in Table I.

b. Feed system. Figure 1 shows a flow diagram of the feed system. Two feedstreams of differing compositions were employed; these were alternatively interchanged by four, electrically operated high-speed solenoid switching valves (Automatic Switch Co.). In cycled experiments, the amplitude is determined by the concentration difference between the two feedstreams, and the frequency by the valve switching frequency. An electronic clock was used to vary the switching frequency in the range of interest (ranging from single steps to about 10 Hz).

The reactor was preceded by a preheater which had a sufficiently small mixing volume so that it did not contribute significantly to the dispersion of the feedstream pulses.

Steady streams of H_2O vapor or SO_2 could be introduced before the preheater to simulate automobile exhaust. When H_2O was added to the feedstream (we used a liquid chromatographic pump), it was condensed out before the gas entered the vent line. However, no water- or SO_2-containing feedstreams were used in the examples shown in this paper.

The gas-phase composition before or after the reactor was monitored by mass spectroscopy; although no such data are shown here, experience indicated the necessity of a fast-response inlet system.

c. Fast-response IR spectrophotometer. A Perkin-Elmer Model 180 infrared spectrophotometer was modified for these experiments to provide the necessary time response. The modifications included the replacement of the original chopper by a high-speed variety (Laser Precision Corporation, 15-1000 Hz), and redirecting the infrared beam so that it was focused onto a

Table I. Properties of the Catalyst Disc

Pt (wt %)	1.0	BET surface area (m^2/g)	100
Pt dispersion (%)*	64	ρ_s (g/cm^3)	3.67
a $(cm^2$ Pt/cm^3 pellet)	25 276	ρ_p (g/cm^3)	1.52
V_{macro} (cm^3/g)	0.040	D_{eff} (cm^2/s)**	0.0246
V_{micro} (cm^3/g)	0.346	ε_p	0.586
$\bar{r}_{macro}$ (Å)	211 150	Diameter (cm)	2.0
$\bar{r}_{micro}$ (Å)	130	Thickness (cm)	0.017

*Determined by a CO flow-chemisorption technique.

**Computed from the random pore model of Wakao and Smith (22) and given here at 101.3 kPa and 723 K.

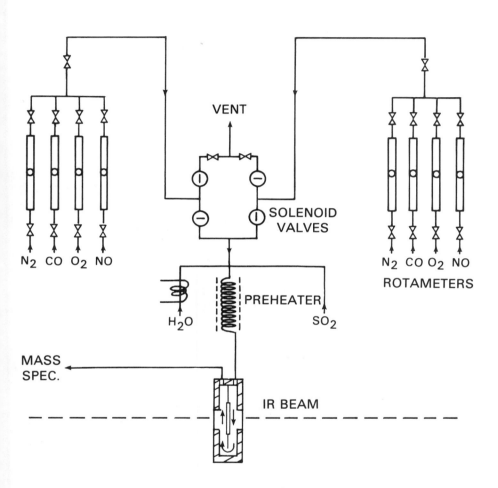

Figure 1.   Schematics of the IR reactor system.

solid-state, high-speed detector (Hg-Cd-telluride, Santa Barbara
Research). The detector was cooled by liquid N_2, and its listed
response time was on the order of 10^{-6} s.

The infrared spectrometer was operated in the single-beam
mode. The detector signal was passed through a lock-in amplifier (Princeton Applied Research, 2 Hz–210 kHz) and the amplified signal was monitored by a storage oscilloscope. The photo-
graphed patterns were then digitized and appropriately processed.
The timer section of the oscilloscope was used to trigger the
switching valves in single-step experiments.

The half-scale response time of the modified infrared
electronics was 0.02 s, as opposed to the half-scale response
time of about 0.5 s for the infrared spectrometer before the
above-described modifications.

d.   Cell response and mixing characteristics. The gas-
phase response and mixing characteristics of the reactor cell
were analyzed by observing the responses of the cell to step
changes in feedstream composition under nonreactive conditions.
About 10 vol % CO in N_2 was employed as the tracer, and the
gas-phase CO band (at 2170 cm^{-1}) was monitored. Analysis of the
step responses indicated reasonably good mixing behavior in the
cell; for a step increase (i.e., $N_2 \rightarrow CO$), a plot of $\ln(1-c_\infty/$
$c_{\infty,in})^{-1}$ vs t yielded a straight line, as shown in Figure 2.
From the slope of the straight line, the effective mixing cell
volume was calculated to be 30.1 cm^3, with a 50% relaxation time
of about 0.08 s. Similar mixing characteristics were observed
following a step decrease (i.e., $CO \rightarrow N_2$), giving an effective
mixing cell volume of 31.8 cm^3 and a 50% relaxation time of
0.09 s. Since these response times of the reactor are not much
faster than the time scale of the adsorption process (a half-
scale relaxation time of about 0.2 s), the transients of the
reactor cell were included in our analysis. For our simulations, the mixing cell volume was taken to be 31 cm^3.

Theory of Transient, Diffusion-Coupled CO Chemisorption

The balance equation for CO in the intrapellet gas phase is

$$\varepsilon_p \frac{\partial c}{\partial t} = D_{eff} \frac{\partial^2 c}{\partial x^2} - a [\tilde{R}_a - \tilde{R}_d] \tag{1}$$

where $\tilde{R}_a$ and $\tilde{R}_d$ represent the rates of CO adsorption and desorption per unit surface area of Pt, respectively. The boundary
conditions for Equation (1) are

$$\frac{\partial c}{\partial x} (0,t) = 0 \tag{2}$$

$$D_{eff} \frac{\partial c}{\partial x} (L,t) = k_m [c_\infty(t) - c(L,t)] \tag{3}$$

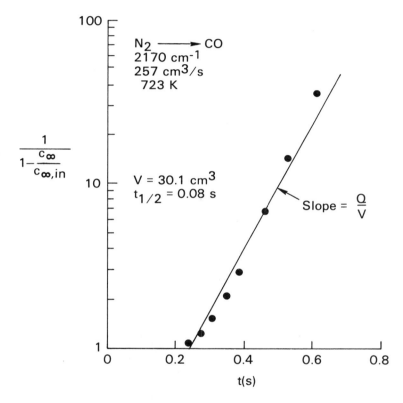

Figure 2.   Mixing behavior of the reactor cell: gas phase response to a step increase in inlet CO concentration.

where L is the half-thickness of the catalyst disc. The bulk gas-phase CO concentration in the reactor, c_∞, varies with time, and its time dependency will be described below.

The conservation of adsorbed CO on the Pt surface can be expressed, in terms of its fractional surface coverage Θ, as

$$N_s \frac{\partial \Theta}{\partial t} = \tilde{R}_a - \tilde{R}_d \tag{4}$$

where N_s denotes the surface concentration of CO corresponding to a complete monolayer coverage of the Pt surface. For our computations, N_s was taken to be 2×10^{-9} mol CO/cm^2 Pt (1.2×10^{15} CO molecules/cm^2 Pt), based on the assumption of 1:1 stoichiometry between Pt surface atoms and CO molecules.

The dynamics of the infrared cell can be described by

$$V \frac{dc_\infty}{dt} = Q [c_{\infty,in} - c_\infty(t)] - D_{eff} A \frac{\partial c}{\partial x} (L,t) \tag{5}$$

In our experiments, the CO concentration at the reactor inlet, $c_{\infty,in}$, was perturbed in a stepwise fashion at t=0.

The initial conditions for Equations (1), (4), and (5) are given by

$$\left. \begin{array}{l} c \ (x,0) = c_o \\ \Theta \ (x,0) = \Theta_o \\ c_\infty(0) \ \ = c_{\infty,o} \end{array} \right\} \tag{6}$$

For the simulation results reported here, $c_o = \Theta_o = c_{\infty,o} = 0$ for CO adsorption, and $c_o = c_{\infty,o} = 1$ vol % and $\Theta_o = 0.78$ for CO desorption. The initial coverage (Θ_o) for the desorption experiments was obtained by calibrating the absorbances in terms of CO coverage, as will be explained later.

The analysis of the system's dynamic behavior requires solving Equations (1) to (6) simultaneously. The rate expressions for $\tilde{R}_a$ and $\tilde{R}_d$ which appear in these equations will be discussed briefly in the following paragraphs.

The rate of adsorption of CO is equal to the product of the flux of CO molecules striking the surface, F, and their sticking probability, S. That is,

$$\tilde{R}_a = F S \tag{7}$$

where F can be written from the kinetic theory of gases (e.g., 23) as

$$F = \sqrt{\frac{R T}{2 \pi M}} \ c \tag{8}$$

and, assuming that the CO adsorption rate is proportional to the fraction of vacant sites,

$$S = S_o (1-\Theta) \tag{9}$$

It should be noted that Equation (9) is not strictly valid
for CO/Pt systems; experiments show that the rate of CO adsorption remains essentially constant as Θ increases up to about 0.25
to 0.5 before the rate decreases with a further increase in Θ (8,
24, 25). However, this simple relationship is adopted here
because the essential features of the system are not expected to
be lost by this approximation. The initial sticking probability
for CO, S_o, remains nearly constant up to about 450 K, but at
higher temperatures it decreases appreciably with temperature (4,
6). In this study S_o (at 723 K) was taken to be 0.025 (4).
    The rate of desorption of an adsorbed species from a surface
is usually expressed in the form

$$\tilde{R}_d = \nu \, c_s^n \, \exp \, (-E_d(\Theta)/RT) \tag{10}$$

where c_s is the surface concentration of the adsorbate (mol/cm^2
Pt surface). In view of the nondissociative adsorption of CO,
the order of desorption n and the pre-exponential factor ν were
taken to be 1 and 10^{13} s^{-1}, respectively (26, 27). A coverage-
dependent activation energy for desorption is assumed because the
desorption energy has often been observed to decrease with coverage for CO/Pt systems (e.g., 8). Moreover, this coverage
dependence has been successfully used in the analysis of thermal
desorption data (7, 24, 28, 29). The decrease in the desorption
energy with coverage is related physically to the mutual interactions (such as long-range repulsive interactions) between the
adsorbed CO molecules (8).
    For the present analysis, it is assumed that the activation
energy for CO desorption decreases linearly with CO coverage.
That is,

$$E_d = E_{do} - \alpha \, \Theta \tag{11}$$

where the desorption energy at the low-coverage limit, E_{do}, was
taken to be 144.4 kJ/mol (34.5 kcal/mol). This desorption energy
is compatible with the literature values reported for single-
crystal Pt surfaces (8, 24, 30, 31), and for Pt ribbons and wires
(3-6). The parameter α, which describes the variation of the
desorption energy with coverage, was chosen to be 27.2 kJ/mol
(6.5 kcal/mol) (7).
    Implicit in the use of Equation (10) is the assumption that
desorption occurs from the surface populated by one type of
adsorbed species. This single-state model will be shown to be
adequate to describe the dynamic behavior of CO desorption at the
high temperature (723 K) considered here. The analysis of the
more general case of multi-state desorption can be found in
Donnelly et al. (32) and Winterbottom (3).
    The determination of the external mass transfer coefficient
of CO, k_m, (see Equation 3) deserves brief comments. Since the
complex geometry and flow characteristics in the reactor cell
precluded a reliable estimation of k_m based on correlations given
in the literature, the CO oxidation activities of the catalyst

disc, measured independently under mass transfer-limited conditions, were used to determine k_m. Figure 3 shows the CO conversion performance of the reactor as a function of temperature in a
large excess of O_2 (i.e., 1 vol % CO and 5 vol % O_2). It can be
seen that the CO conversion remains essentially constant (about
60%) above 650 K, indicating that the reaction rate in this
regime is limited by mass transfer from the bulk gas to the
catalyst's external surface. Similar conversion data, including
the same asymptotic conversion level, were obtained with a feedstream composition of 0.5 vol % CO and 5 vol % O_2. In the completely mass transfer-limited regime, the steady-state mass
balance equation for the reactor cell can be written as

$$Q \, (c_{\infty,in} - c_{\infty}) \ = \ k_m \, A \, c_{\infty} \tag{12}$$

or, in terms of the fractional CO conversion ϕ,

$$k_m \ = \ \frac{Q}{A} \, (\frac{\phi}{1-\phi}) \tag{13}$$

Using Equation (13), the external mass transfer coefficient at
723 K was calculated to be 60 cm/s. Since the reactor operating
conditions at this temperature (723 K, slightly above atmospheric
pressure, 247 cm^3/s) were very similar to those of our transient
chemisorption experiments, the external mass transfer coefficient
calculated above was used for the simulations.

The transmittance observed at 2070 cm^{-1} was converted into
absorbance ($A_{CO} = \ln(1/T_{CO})$), since the latter is a direct measure
of the surface concentration of CO. In order to quantify the
absorbances in terms of surface coverages, the steady-state
intensities of the Pt-CO band were measured at several different
gas-phase CO concentrations. The results (absorbance vs CO
concentration) are shown in Figure 4a. The same data, when
plotted in the plane of c/A_{CO} vs c, yield a straight line (see
Figure 4b), indicating that these adsorption equilibrium data can
be well approximated by a Langmuir isotherm. From the slope of
this straight line, the absorbance corresponding to a monolayer
coverage of CO, A_{sat}, was calculated to be 0.473.

Once A_{sat} is determined, the surface coverage of CO can be
obtained from the relationship $\theta = A_{CO}/A_{sat}$. This relationship
implies that the extinction coefficient of adsorbed CO is independent of CO coverage. While this aspect was not explored here,
the agreement of the model's predictions with the experimental
data (as will be shown later) indicates that this assumption may
not interfere significantly with the essential features of the
system.

For the rate expressions $\tilde{R}_a$ and $\tilde{R}_d$ given above, Equations
(1) to (6) were solved numerically using the PDEPACK routine of
Madsen and Sincovec (33). The computational results as well as
their comparison with the experimental data will be discussed in
the next section.

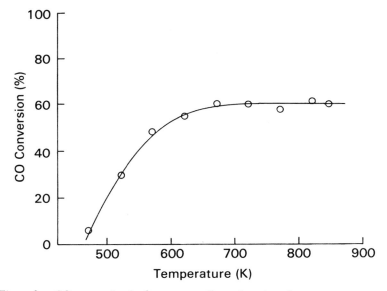

Figure 3. CO conversion in the reactor cell as a function of temperature, measured to determine the external mass transfer coefficient of CO. Conditions: 1% CO, 5% O_2, 93.3 cm^3/s at 273 K and 101.3 kPa.

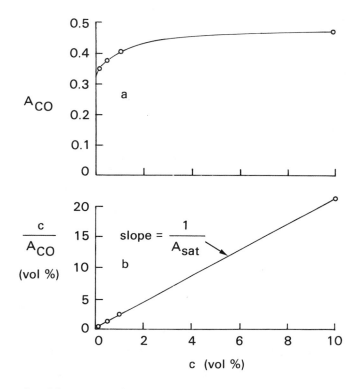

Figure 4.   Adsorption isotherm for the CO/Pt system at 723 K and 121.7 kPa, plotted in two different forms. In the lower curve, $A_{sat} = 0.473$.

Results and Discussion

a.   CO chemisorption studies.   The step-change experiments
for the adsorption and desorption of CO were conducted at a
pellet temperature of 723 K and at a reactor pressure of 121.7
kPa (slightly above atmospheric pressure). Such a high temperature was chosen in order to approximate the operating temperature
of a fully warmed-up catalytic converter. The gas flow rate was
257 cm^3/s at our experimental conditions. Before the infrared
experiments, the Pt-alumina catalyst disc was pretreated with
0.7 vol % O_2 (in N_2), with N_2, with 1 vol % H_2 (in N_2), and then
with N_2, each step being at 723 K for 5 min. The adsorption of
CO was then studied by suddenly directing a feedstream of 1 vol %
CO in N_2 through the reactor (in place of the N_2 stream) and
monitoring the resulting time variations of the transmittance at
2070 cm^{-1}, the well-known Pt-CO stretch frequency. After equili-
bration, the step response of CO desorption was measured by
switching the feed back to N_2.
    It is interesting to note that at this high temperature the
frequency of the Pt-CO band did not shift appreciably as the CO
coverage varied, in contrast to observations made on a (111)-
oriented platinum ribbon (6) or on a polycrystalline Pt surface
(34) at or below room temperature.
    As will be shown later, the surface coverages of CO vary
with distance into the pellet during CO adsorption and desorption, as a result of intrapellet diffusion resistances. However,
the infrared beam monitors the entire pellet, and thus the resulting absorption band reflects the average surface concentration of CO across the pellet's depth. Therefore, for the purpose
of direct comparison between theory and experiment, the integral-
averaged CO coverage in the pellet

$$\Theta_{av}(t) = \frac{1}{L} \int_o^L \Theta(x,t)\ dx \qquad (14)$$

was computed from the mathematical model.
    Figure 5 shows the integral-averaged surface coverage of CO
in the catalyst pellet as a function of time during the CO adsorption at 723 K. In this experiment, the reactor inlet concentration was switched from N_2 to 1 vol % CO (in N_2) at t=0,
while maintaining the total flow rate and the reactor pressure
essentially unperturbed. The measured values (solid line in
Figure 5) were obtained based on the relationship $\Theta_{av}=A_{CO}/A_{sat}$
($A_{sat}=0.473$). The asymptotic CO coverage reached after the
relaxation of the step change was calibrated to be 78% of a
complete monolayer coverage. The computational result shown in
the dotted line of Figure 5 is in very good agreement with the
experimental data.
    Figure 6 shows the measured and calculated time variations
of the integral-averaged CO coverage following the step change

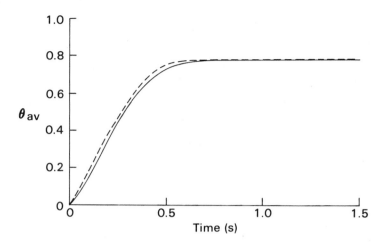

Figure 5. Integral-averaged CO coverage as a function of time during CO adsorption. Key: ———, measured; and – – –, calculated.

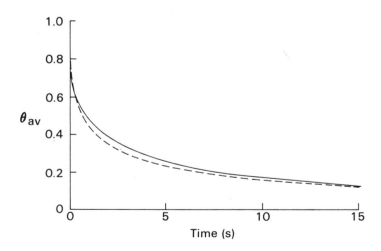

Figure 6. Integral-averaged CO coverage as a function of time during CO desorption. Key: ———, measured; and – – –, calculated.

from 1 vol % CO to N_2 (i.e., during CO desorption at 723 K).
Here again, good agreement between theory and experiment was
obtained. Notice that the CO desorption occurred much more
slowly than the CO adsorption; that is, a ten-fold difference in
the 50% relaxation time (2 s vs 0.2 s) was observed.

In view of the possible presence of diffusion resistances in
the catalyst pellet, it is of interest to examine the intrapellet
concentration profiles established during transient CO adsorption
and desorption. Figure 7 shows the calculated intrapellet concentration profiles for CO adsorption at various times. It can
be seen that significant gas-phase concentration gradients (see
solid lines) are established inside the pellet at the early
stages of the transient adsorption process. Also shown in Figure
7 are the corresponding profiles for the fractional surface
coverage of CO (dotted lines).

Figure 8 shows how the intrapellet concentration profiles
vary with time during the course of CO desorption. Both the gas-
phase (solid lines) and surface (dotted lines) CO concentration
profiles exhibit relatively mild gradients inside the pellet, in
contrast to the steep profiles established during the adsorption
process. This can be attributed to the fact that the intrinsic
rate of desorption is slower than that of adsorption.

The importance of including mass transport resistances (both
internal and external) in the model can be best examined by
comparing the exact solution of the model with the predicted
responses obtained based on the assumption of negligible mass
transport resistances. As Figure 9 shows, the step response of
the integral-averaged CO coverage during the adsorption is affected only slightly by the external mass transfer resistance
(compare Curves A and B); however, when both internal and external mass transport resistances are assumed to be negligible,
the model predicts a considerably faster response (compare
Curves A and C). These results demonstrate the importance of
accounting for intrapellet diffusion resistances in the analysis
of transient chemisorption over supported catalysts. Similar
effects are observed for CO desorption, as shown in Figure 10.

It is interesting to note that, although the intrinsic rate
of desorption is slower than that of adsorption, both rates were
found to be sufficiently fast under our experimental conditions
so that the adsorption-desorption process on the Pt surface can
be assumed to rapidly equilibrate at all times; that is, even a
ten-fold increase in both the adsorption and desorption rate
constants (while keeping their ratio constant) did not significantly change the predicted step responses. With the assumption of chemisorption equilibrium, Equations (1) and (4) can be
combined into the form (35)

$$[\varepsilon_p + N_s \, a \, (\frac{\partial\theta}{\partial c})] \, \frac{\partial c}{\partial t} = D_{eff} \, \frac{\partial^2 c}{\partial x^2} \qquad (15)$$

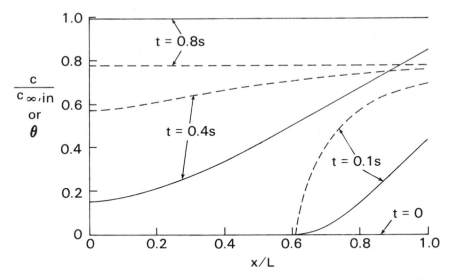

Figure 7.   Computed time variation of intrapellet concentration profiles during CO adsorption. Key: ——, gas phase; and – – –, surface.

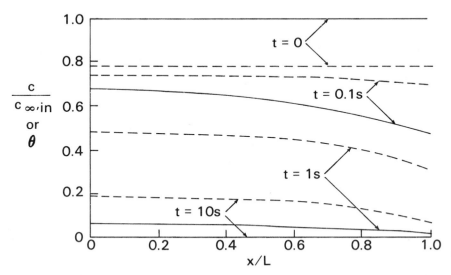

Figure 8.   Computed time variation of intrapellet concentration profiles during CO desorption. Key: ——, gas phase; and – – –, surface.

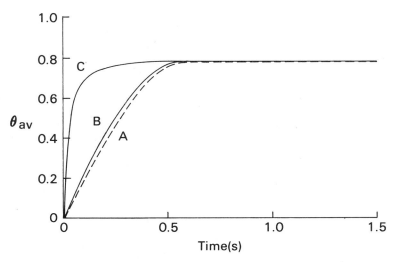

Figure 9.   Effects of internal and external transport resistances on the computed step-response of CO adsorption. Curve A corresponds to our experimental conditions. Key: A, k_m = 60 cm/s, D_{eff} = 0.0246 cm^2/s; B, $k_m \to \infty$, D_{eff} = 0.246 cm^2/s; and C, $k_m \to \infty$, $D_{eff} \to \infty$.

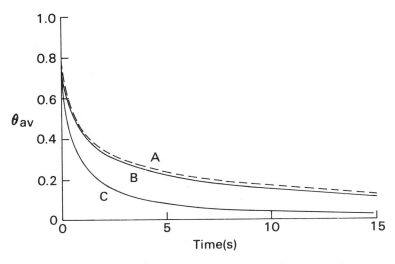

Figure 10.   Effects of internal and external transport resistances on the computed step-response of CO desorption. Curve A corresponds to our experimental conditions. Key: A, k_m = 60 cm/s, D_{eff} = 0.0246 cm^2/s; B, $k_m \to \infty$, D_{eff} = 0.0246 cm^2/s; and C, $k_m \to \infty$, $D_{eff} \to \infty$.

where $(\partial\Theta/\partial c)$ represents the slope of the adsorption isotherm. The quantity $[N_s \ a \ (\partial\Theta/\partial c)]$ in Equation (15) accounts for the capacity of the catalytic surface for the chemisorption of CO. For porous catalyst pellets with practical loadings, this quantity is typically much larger than the pellet void fraction ε_p, indicating that the dynamic behavior of supported catalysts is dominated by the relaxation of surface phenomena (e.g., 35, 36). This implies that a quasi-static approximation for Equation (1) (i.e., $\varepsilon_p \frac{\partial c}{\partial t} = 0$) can often be safely invoked in the transient modeling of porous catalyst pellets. The calculations showed that the quasi-static approximation is indeed valid in our case; the model predicted virtually the same step responses, even when the value of ε_p was reduced by a factor of 10.

It follows directly from Equation (15) that the characteristic response time for the intrapellet gas-phase concentration during the adsorption-desorption process is approximated by $[(L^2/D_{eff}) \ N_s \ a \ (\partial\Theta/\partial c)]$. (Note that this characteristic response time represents the product of the characteristic time for intrapellet diffusion, L^2/D_{eff}, and the surface capacity for CO chemisorption, $N_s \ a \ (\partial\Theta/\partial c)$.) The estimation of the response time is complicated by the fact that the slope of the adsorption isotherm $(\partial\Theta/\partial c)$ varies with gas-phase CO concentration; however, when its average value over the CO concentration range of 0 to 1 vol % was taken to be 5×10^6 cm^3/mol (= 1 $(vol \ \%)^{-1}$; see Figure 4a), the characteristic response time was calculated to be 0.75 s. Inspection of the time variation of the intrapellet gas-phase concentrations shown in Figures 7 and 8 (solid lines) lends support to our estimated response time; that is, the intrapellet gas-phase concentration relaxes on the time scale of 1 s during both the adsorption and desorption processes.

In contrast to the intrapellet gas-phase concentration, the surface concentration of CO in the pellet relaxes at significantly different rates, depending on the direction of the step changes. As shown earlier, the surface concentration of CO decays much more slowly during CO desorption than it increases during CO adsorption (compare Figures 5 and 6). This is, however, merely due to the fact that close to adsorption equilibrium, the surface coverage of CO can be high even at low gas-phase CO concentrations, as shown in Figure 4a.

b. Reactive experiments. In addition to the chemisorption studies described above, we also conducted transient infrared experiments under reaction conditions (i.e., CO + O_2 over Pt-alumina). The results of these reactive experiments, though not yet analyzed in quantitative detail, will be nevertheless shown here (in the transmittance mode) because they illustrate some interesting features of the transient response of the $CO/O_2/Pt$ system.

Figure 11 shows how O_2 pretreatment of the catalyst influences the step response during CO adsorption. Prior to this experiment, the catalyst was first pretreated in 0.7 vol % O_2 (in N_2) for 5 min and then flushed with N_2 for 5 min (all at 723 K). After this N_2 flush, the catalyst was suddenly exposed to a step flow of 1 vol % CO (in N_2). As shown in Figure 11, the growth of the 2070 cm^{-1} Pt-CO band is significantly delayed after O_2 pretreatment (Curve B), indicating that at the early stages of the step change, the CO on the Pt surface is rapidly consumed by the strongly chemisorbed oxygen.

The rate of CO removal from the Pt surface is also affected by the presence of O_2 in the gas phase, as demonstrated in Figure 12. In this experiment the catalyst, initially in equilibrium with 1 vol % CO (in N_2), was suddenly exposed to a feedstream of 0.7 vol % O_2 (in N_2). It can be seen from Figure 12 that the Pt-CO band decays much faster in O_2 (Curve B) than in N_2 (Curve A for reference). This indicates that the surface reaction between CO and oxygen is faster than the rate of CO desorption.

Figure 13 shows the time variation of the surface concentration of CO (on the transmittance scale) during feedstream composition cycling. Cycling between two feedstreams, one oxidizing (1 vol % CO, 1.03 vol % O_2) and the other reducing (1 vol % CO, 0.23 vol % O_2), was accomplished by four fast-acting solenoid valves at a switching frequency of 1 Hz. In the oxidizing feedstream, the Pt surface was found to be essentially free of CO, while a significantly higher CO coverage was observed over Pt in equilibrium with the reducing feedstream. It is interesting to note that, as Figure 13 shows, the surface concentration of CO assumed the same stationary pattern after several transitory cycles, regardless of whether the catalyst was initially stabilized in an oxidizing or in a reducing feedstream.

Concluding Remarks

A fast-response infrared spectroscopic reactor system has been described which is capable of operating at high temperatures (e.g., 450-500°C). The infrared reactor system was successfully used to monitor the response of the surface concentration of CO to step changes or oscillations in the feedstream composition, under both reactive and nonreactive conditions.

The dynamics of high-temperature CO adsorption and desorption over Pt-alumina was analyzed in detail using a transient mathematical model. The model combined the mechanism of CO adsorption and desorption (established from ultrahigh-vacuum studies over single-crystal or polycrystalline Pt surfaces) with extra- and intrapellet transport resistances. The numerical values of the parameters which characterize the surface processes were taken from the literature of clean surface studies;

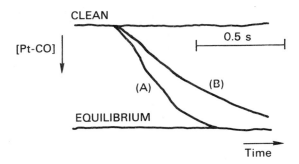

Figure 11.   Effect of O_2 pretreatment on the rate of CO adsorption (transmittance mode) at 723 K, 257 cm^3/s, and 2,070 cm^{-1}.   Key: A, prereduced Pt; and B, Pt preexposed to O_2 and flushed with N_2.

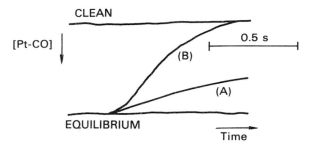

Figure 12.   Effect of O_2 on the rate of CO removal (transmittance mode) at 723 K, 257 cm^3/s, and 2,070 cm^{-1}.   Key: A, CO removal by N_2; and B, CO removal by 1% O_2.

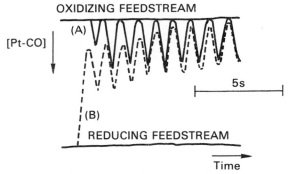

Figure 13.   Surface transients over Pt-alumina, starting with an oxidizing feedstream (A) or with a reducing feedstream (B) (transmittance mode) at 723 K, 257 cm^3/s, 2,070 cm^{-1}, and 1 Hz switching frequency.

we found good agreement between the measured and calculated step
responses, indicating that, at least in this case, it is possible to apply surface chemistry information to supported catalysts
operating at high temperature and pressure (450°C, slightly
above atmospheric pressure).

Parametric sensitivity analysis showed that for nonreactive
systems, the adsorption equilibrium assumption can be safely invoked for transient CO adsorption and desorption, and that intrapellet diffusion resistances have a strong influence on the time
scale of the transients (they tend to slow down the responses).
The latter observation has important implications in the analysis
of transient adsorption and desorption over supported catalysts;
that is, the results of transient chemisorption studies should be
viewed with caution, if the effects of intrapellet diffusion resistances are not properly accounted for.

Nomenclature

a	=	local Pt surface area, cm^2 Pt/cm^3 pellet
A	=	total external surface area of the catalyst pellet, cm^2
A_{CO}	=	absorbance for adsorbed CO
A_{sat}	=	absorbance corresponding a complete monolayer coverage of CO
c	=	CO concentration in the intrapellet gas phase, mol/cm^3 or vol %
c_o	=	initial CO concentration in the intrapellet gas phase, mol/cm^3
c_s	=	surface concentration of CO, mol/cm^2 Pt
c_∞	=	CO concentration in the bulk gas phase of the reactor cell, mol/cm^3
$c_{\infty,o}$	=	initial CO concentration in the reactor cell, mol/cm^3
$c_{\infty,in}$	=	CO concentration at the inlet of the reactor cell, mol/cm^3
$\cent$	=	fractional conversion of CO
D_{eff}	=	effective diffusivity of CO in the catalyst pellet, cm^2/s

E_d = activation energy for desorption, kJ/mol

E_{do} = activation energy for desorption extrapolated to zero surface coverage of CO, kJ/mol

F = flux of CO molecules striking the surface, mol/$(s \cdot cm^2$ Pt)

k_m = external mass transfer coefficient of CO, cm/s

L = half-thickness of the catalyst pellet, cm

M = molecular weight of CO, g/mol

N_s = saturation CO concentration over the active sites, mol/cm^2 Pt

Q = gas volumetric flow rate, cm^3/s

R = gas constant, 8.3144×10^7 g cm^2/$(s^2 \cdot mol \cdot K)$ or 8.32×10^{-3} kJ/$(mol \cdot K)$

$\tilde{R}_a$ = rate of CO adsorption, mol/$(s \cdot cm^2$ Pt)

$\tilde{R}_d$ = rate of CO desorption, mol/$(s \cdot cm^2$ Pt)

$\bar{r}_{macro\ or\ micro}$ = integral-averaged pore radii, Å

S = sticking probability for CO

S_o = initial sticking probability for CO

T = temperature, K or °C

T_{CO} = transmittance for adsorbed CO

t = time, s

V = volume of the reactor cell, cm^3

$V_{macro\ or\ micro}$ = pellet macro- or micropore volumes, cm^3/g

x = distance from the pellet center, cm

Greek Letters

α = parameter describing the variation of the desorption energy with fractional surface coverage (see Equation 11), kJ/mol

ε_p	=	pellet void fraction
Θ	=	fractional surface coverage of CO
Θ_{av}	=	integral-averaged fractional surface coverage of CO in the pellet
Θ_o	=	initial CO coverage
ν	=	pre-exponential factor for the rate of desorption, s^{-1}
ρ_p	=	pellet density, g/cm^3 pellet
ρ_s	=	pellet solid density, g/cm^3 solid

Acknowledgments

E. J. Shinouskis built the infrared reactor cell and conducted the infrared reactor experiments. The authors are indebted to J. A. Sell for the modifications of the infrared spectrophotometer, and to Professor A. T. Bell for sharing his experience with infrared reactor construction. The CO conversion data of Figure 3 were obtained by J. E. Carpenter.

Literature Cited

1.   Langmuir, I. Trans. Faraday Soc. 1922, 17, 621.

2.   Morgan, A. E.; Somorjai, G. A. J. Chem. Phys. 1969, 51, 3309.

3.   Winterbottom, W. L. Surface Sci. 1973, 37, 195.

4.   Nishiyama, Y.; Wise, H. J. Catal. 1974, 32, 50.

5.   Collins, D. M.; Lee, J. B.; Spicer, W. E. Surface Sci. 1976, 55, 389.

6.   Shigeishi, R. A.; King, D. A. Surface Sci. 1976, 58, 379.

7.   McCabe, R. W.; Schmidt, L. D. Surface Sci. 1977, 65, 189.

8.   Ertl, G.; Neumann, M.; Streit, K. M. Surface Sci. 1977, 64, 393.

9.   Cormack, D.; Moss, R. L. J. Catal. 1969, 13, 1.

10. Bain, F. T.; Jackson, S. D.; Thomson, S. J.; Webb, G.; Willocks, E. J. Chem. Soc. Faraday Trans. I 1976, 72, 2516.

11. Foger, K.; Anderson, J. R. Appl. Surface Sci. 1979, 2, 335.

12. Delgass, W. N.; Haller, G. L.; Kellerman, R.; Lunsford, J. H. "Spectroscopy in Heterogeneous Catalysis," Academic Press: New York, 1979.

13. Eischens, R. P.; Francis, S. A.; Pliskin, W. A. J. Phys. Chem. 1956, 60, 194.

14. Heyne, H.; Tompkins, F. C. Proc. Roy. Soc. (London) 1966, A292, 460.

15. Hegedus, L. L.; Chang, C. C.; McEwen, D. J.; Sloan, E. M. Ind. Eng. Chem. Fundam. 1980, 19, 367.

16. Kobayashi, H.; Kobayashi, M. Catal. Rev.-Sci. Eng. 1974, 10, 139.

17. Bennett, C. O. Catal. Rev.-Sci. Eng. 1976, 13, 121.

18. Tamaru, K. "Dynamic Heterogeneous Catalysis," Academic Press: New York, 1978.

19. Hegedus, L. L.; Gumbleton, J. J. ChemTech 1980, 10, 630.

20. Hegedus, L. L.; Summers, J. C.; Schlatter, J. C.; Baron, K. J. Catal. 1979, 56, 321.

21. Bell, A. T., personal communications (1980).

22. Smith, J. M. "Chemical Engineering Kinetics," McGraw-Hill: New York, 1970.

23. Somorjai, G. A. "Principles of Surface Chemistry," Prentice-Hall: Englewood Cliffs, New Jersey, 1972.

24. McCabe, R. W.; Schmidt, L. D. Surface Sci. 1977, 66, 101.

25. Weinberg, W. H.; Comrie, C. M. J. Catal. 1976, 41, 489.

26. Menzel, D. "Desorption Phenomena," in "Topics in Applied Physics," Vol. 4; Gomer R., Ed.; Springer-Verlag: New York, 1975.

27.   Baetzold, R. C.; Somorjai, G. A. J. Catal. 1976, 45, 94.

28.   Falconer, J. L.; Madix, R. J. J. Catal. 1977, 48, 262.

29.   Taylor, J. L.; Weinberg, W. H. Surface Sci. 1978, 78, 259.

30.   Bonzel, H. P.; Burton, J. J. Surface Sci. 1975, 52, 223.

31.   Lambert, R. M.; Comrie, C. M. Surface Sci. 1974, 46, 61.

32.   Donnelly, R. G.; Modell, M.; Baddour, R. F. J. Catal.
       1978, 52, 239.

33.   Madsen, N. K.; Sincovec, R. F. "PDEPACK: Partial Differen-
       tial Equations Package," Scientific Computing Consulting
       Services, Livermore, California, 1975.

34.   Hoffmann, F. M.,; Bradshaw, A. M. J. Catal. 1976, 44, 328.

35.   Oh, S. H.; Baron, K.; Cavendish, J. C.; Hegedus, L. L.
       Paper No. 38, ACS Symposium Series, No. 65, "Chemical
       Reaction Engineering-Houston," ACS: Washington, D.C.,
       1978.

36.   Sheintuch, M.; Schmitz, R. A. Catal. Rev.-Sci. Eng. 1977,
       15, 107.

RECEIVED July 28, 1981.

Nitric Oxide Reduction by Hydrogen over Rhodium Using Transient Response Techniques

BRUCE J. SAVATSKY and ALEXIS T. BELL

University of California, Department of Chemical Engineering, Berkeley, CA 94720

The catalyzed reduction of nitric oxide over rhodium is of considerable interest in view of extensive research (1-7) showing that catalytic convertors containing rhodium are particularly effective for controlling the emission of nitric oxide from automobiles. While it has been established that carbon monoxide and hydrogen are the primary agents participating in NO reduction, little is known thus far concerning the mechanism and kinetics of the reduction process. Studies by Kobylinski and Taylor (1) with Rh/Al_2O_3 have shown that H_2 is a more effective reducing agent than CO, as evidenced by the fact that the temperature required to achieve a given degree of conversion is lower using H_2 rather than CO as the reducing agent. It was also demonstrated that even when H_2 and CO are combined, H_2 is still the preferred reducing agent. A more detailed investigation of NO reduction by H_2 has been reported by Yao et al. (6). These authors noted that the global activation energy for NO reduction decreased from 14.7 to 8.8 kcal/ mole as the Rh loading increased from 0.34 to 12.4%. Nitrogen and N_2O were observed as the major products with N_2O being the dominant product under all conditions studied. It was also noted that the ratio of N_2/N_2O was not very much affected by either the reactant concentrations or the catalyst temperature. Although not observed directly, the authors proposed that a surface complex containing two NO molecules and two H atoms could be an important reaction intermediate. More recently, Myers and Bell (8) have investigated the interaction of NO with H_2 over Rh/SiO_2 using temperature programmed reaction (TPR). Their results suggest that NO dissociation to form N and O atoms is the first step of the reaction sequence. Ammonia and H_2O are believed to be formed by the stepwise hydrogenation of the atomic species and the small amount of N_2 observed is formed by the recombination of N atoms.

The present work was undertaken to investigate the applicability of transient response techniques [e.g., ref. (9-11)] for characterizing the mechanism and kinetics of NO reduction. Mass spectrometry was used to trace the dynamics of product formation and in situ infrared spectroscopy was used to

0097-6156/82/0178-0105$09.25/0

observe the state of adsorbed NO. Two types of experiment were
performed. The first involved the reduction of preadsorbed NO
in a constant flow of H_2. The second type of experiment involved
the substitution of ^{15}NO for ^{14}NO, either during steady-state
reduction or just prior to the addition of H_2 to a flowing
stream of NO. It will be shown that the data obtained from
these experiments can be interpreted in terms of a mechanism of
NO deduced from evidence presented in the recent literature.

Experimental

   Catalyst. A 4.4% Rh/SiO_2 catalyst was used for all of the
work presented here. The catalyst was prepared by impregnation
of Davison 70 silica gel with an aqueous solution of $RhCl_3$. The
freshly prepared material was dried and then reduced in flowing
H_2 at 673K for 2 hr. The dispersion of the catalyst was determined to be 25% by H_2 chemisorption.

   Apparatus. Figure 1 shows a schematic of the reactor. The
reactor body was made of stainless steel flanges. To eliminate
the catalytic activity of the reactor, the interior walls were
coated with aluminum and then oxidized to produce an alumina
surface. Calcium fluoride windows were mounted in each half of
the body so that an infrared beam can be passed through the
reactor. These windows were sealed to the stainless steel
flanges by compression between two Graphoil gaskets. The seal
between the two halves of the reactor was made through a copper
gasket, gold plated to eliminate its catalytic activity. The
catalyst, a 15 mm diameter disk, weighing 51 mg, was held in
place between two aluminum rings. The assembled reactor is
compact and has a dead volume of only 1.6 cm^3. The reactor was
heated by two 200W disk heaters (Thermal Circuits, Inc.), placed
over the circular faces, and the catalyst temperature was
measured with a 1/16" stainless-steel sheathed thermocouple, fed
through a port (not shown) in the reactor body.
   The reactor was connected to the balance of the apparatus as
shown in Fig. 2. One of two premixed gas streams could be fed to
the reactor. By switching the indicated valve, a step change in
either reactant or isotope concentration could be achieved. The
effluent from the reactor was sampled through a differentially
pumped inlet system into an EAI 250B quadrupole mass spectrometer.
The signal from the mass spectrometer was then fed to a micro-
processor-based data acquisition system. This unit was designed
to monitor as many as ten preselected mass peaks, as well as the
catalyst temperature and the gas flow rate. The time required for
the collection of one data set depended on the number of mass peaks
tracked and the dwell time on each peak. The shortest cycle time
used in the present work was 0.5 s, during which the intensities
of four mass peaks were read, together with the catalyst temperature and the gas flow rate.

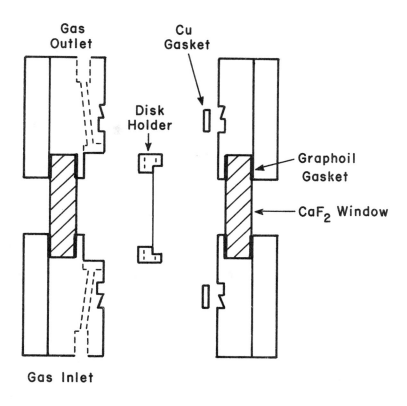

Figure 1.   Schematic of the reactor — IR cell.

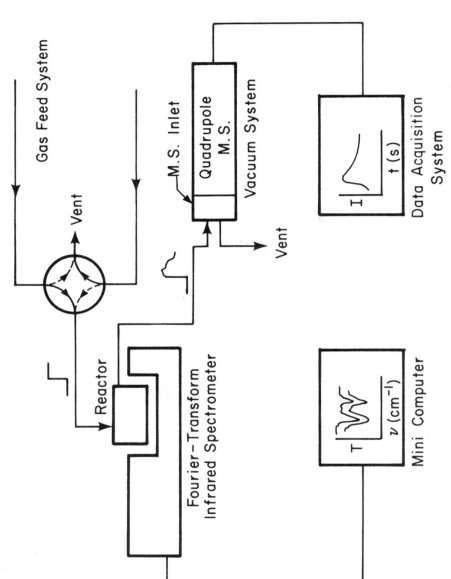

Figure 2.   Schematic of the experimental apparatus.

Transmission infrared spectra of species adsorbed on the catalyst were taken with a Digilab FTS-10M Fourier-transform infrared spectrometer, using a resolution of 4 cm^{-1}. To improve the signal-to-noise ratio, between 10 and 100 interferograms were co-added. Spectra of the catalyst taken following reduction in H_2 were subtracted from spectra taken in the presence of NO to eliminate the spectrum of the support. Because of the very short optical path through the gas in the reactor and the low NO partial pressures used in these studies, the spectrum of gas-phase NO was extremely weak and did not interfere with the observation of the spectrum of adsorbed species.

Results

Reaction of Preadsorbed NO. These experiments were carried out using the following procedure. Prior to each experiment, the catalyst was reduced in H_2 for 5 min at 423 K. The catalyst was then exposed to a 0.28% NO/Ar mixture for 5 to 60 s. It was noted that NO adsorption was accompanied by a partial reduction of the adsorbing NO with adsorbed H_2 left on the catalyst surface from the period of reduction. Infrared spectra taken at the end of the NO adsorption period showed an intense band at 1660-1680 cm^{-1}, the position of the band shifting upscale with the duration of adsorption, and a much weaker band at 1830 cm^{-1}. Based on previous studies of NO adsorption on supported Rh catalysts (12-14) and analogy with the spectra of Rh nitrosyls (15-17), the band at 1660-1680 cm^{-1} can be assigned to a $NO_a^{\delta-}$ structure and the band at 1830 cm^{-1} to a NO_a structure. The reduction of adsorbed NO was initiated by substituting a 10% H_2/Ar mixture for the NO/Ar mixture. While the formation of products began immediately, no change in the catalyst temperature was observed during the course of the reaction.

Figure 3 illustrates a typical series of product responses. Nitrogen and N_2O were formed immediately upon contact of the catalyst with H_2 and the maximum in the signals for these products was observed in about 1.2 s. The hydrogen-containing products, NH_3 and H_2O, also appeared as sharp peaks, but the maximum in the signals for these products occurred at 2.5 s. Changes in the experimental conditions did not alter the qualitative features of the responses. In all cases the peaks for N_2 and N_2O occurred simultaneously and preceded the peaks for NH_3 and H_2O. It was further observed that the reaction temperature, duration of NO exposure, and the H_2 partial pressure during NO reduction affected the intensities of the product peaks and the time delay between the N_2/N_2O peaks and the NH_3/H_2O peaks. To a much lesser degree, the reaction conditions also affected the time at which the N_2 and N_2O signals reached a maximum.

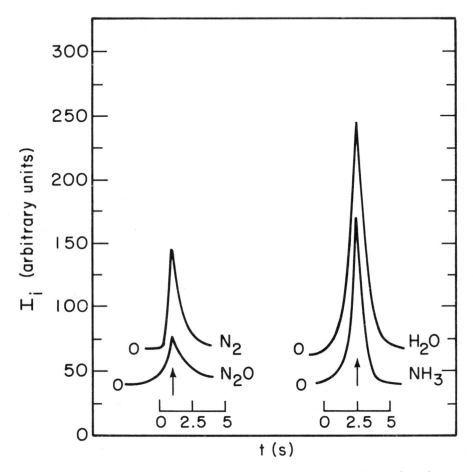

Figure 3.   The transient responses for N_2, N_2O, H_2O, and NH_3, during the reduction of preadsorbed NO. Before reaction, NO was adsorbed for 15 s: $P_{H_2} = 1.0 \times 10^{-1}$ atm, $P_{NO} = 2.8 \times 10^{-3}$ atm, and $T = 423$ K.

Figure 4 shows the effects of NO exposure time on the absorbance of the band at 1660–1680 cm^{-1}, observed prior to reaction, and the maximum intensities of the NH_3, N_2, N_2O, and H_2O signals, observed during reaction. The absorbance of the infrared band provides a measure of the quantity of NO adsorbed on the Rh surface. As may be seen, the coverage by adsorbed NO increases rapidly for NO exposures up to 15 s and then levels off at a plateau. The product maxima show a similar trend, but for NO exposures greater than 20 s, the signals for H_2O, NH_3, and N_2 show a slight decline. Assuming that the maximum signal intensity for each product is proportional to the total amount of product formed, the results presented in Fig. 4 indicate that for NO exposure times greater than 20 s the amount of adsorbed NO undergoing reduction decreases slightly with increasing NO coverage. Such a trend suggests that at high NO coverages, a fraction of the adsorbed NO desorbs and, hence, is not reduced.

The time of exposure to NO had no effect on the time required for the N_2 and N_2O signals to reach a maximum during NO reduction. However, the NO exposure time did affect the time at which the NH_3 and H_2O signals attained a maximum. As shown in Fig. 5, the delay between the maximum for the N_2 peak and the maximum for the NH_3 peak was the same as the delay between the N_2 peak and the H_2O peak. In both cases, as the exposure time increases, the delay increases from 0 to 2.5 s and then levels off for NO exposures of more than 20 s.

Results similar to those shown in Figs. 4 and 5 were also obtained at other temperatures. Increasing the temperature from 398 to 473 K had two principal effects. The first was to produce narrower but more intense product peaks. The second effect was to reduce the delay between the N_2 peak and either the NH_3 or H_2O peak. Thus, for example, for NO exposure times of greater than 20 s, the delay decreased from 7.5 s at 398 K to 0.5 s at 473 K.

The effects of the H_2 partial pressure, used to reduce the adsorbed NO, on the maximum product signal intensities and the times at which the product peaks appeared was explored for a constant NO exposure time of 35 s and a temperature of 423 K. The results in Fig. 6 show that the maximum intensity of each product signal increases slightly as the partial pressure of H_2 is increased. Both the time at which N_2 and N_2O signals reach a maximum and the delay of the NH_3 and H_2O signals are affected by the H_2 partial pressure. Figure 7 show that the time for maximum production of N_2 and N_2O decreases from 3 to 1.5 s as the H_2 partial pressure increases from 0.01 to 0.1 atm and Fig. 8 shows that the time delay for the maximum production of NH_3 and H_2O decreases from 5.5 to 3.0 s.

Isotopic Tracing with ^{15}NO. To trace out the dynamics of nitrogen incorporation into the products under conditions where the surface coverage by adsorbed NO remains constant, experiments

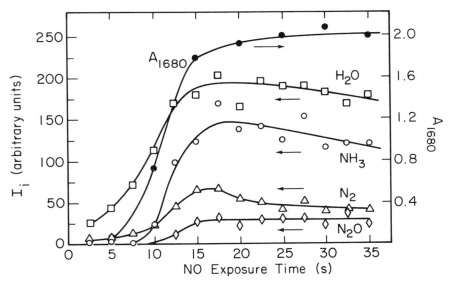

Figure 4. The maximum intensity of the reaction products and the absorbance of the IR band at 1680 cm⁻¹, as a function of NO exposure time during the reduction of preadsorbed NO_2: $P_H = 1.0 \times 10^{-1}$ atm, $P_{NO} = 2.8 \times 10^{-3}$ atm, and $T = 423$ K.

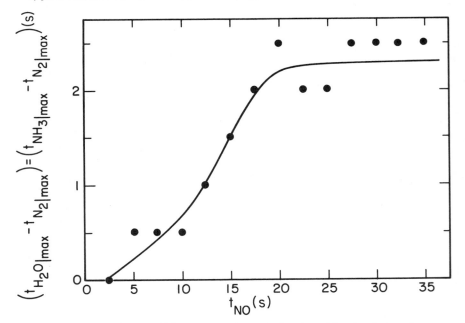

Figure 5. The effect of NO exposure time on the time delay in the maximum production of H_2O and NH_3 relative to N_2, during the reduction of preadsorbed NO: $P_{H_2} = 10 \times 10^{-1}$ atm, $P_{NO} = 2.8 \times 10^{-3}$ atm, and $T = 423$ K.

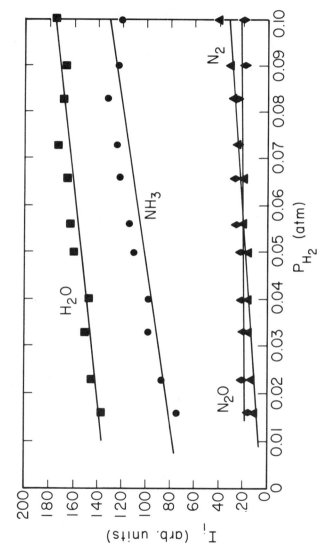

Figure 6. The effect of H_2 partial pressure on the maximum intensities of H_2O, NH_3, N_2O, and N_2, during the reduction of preadsorbed NO: $P_{NO} = 2.8 \times 10^{-3}$ atm, $T = 423$ K, and $t_{NO} = 30$ s.

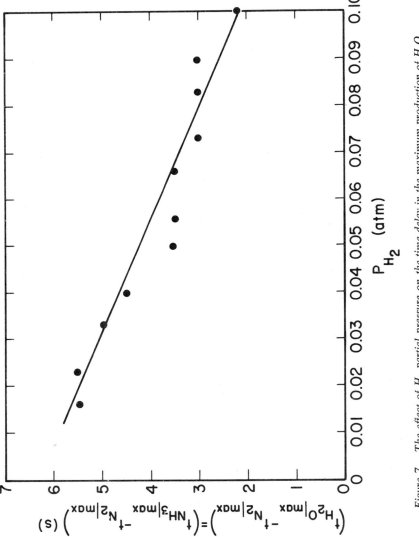

Figure 7. The effect of H_2 partial pressure on the time delay in the maximum production of H_2O and NH_3 relative to N_2, during the reduction of preadsorbed NO: $P_{NO} = 2.8 \times 10^{-3}$ atm, $T = 423$ K, and $t_{NO} = 30$ s.

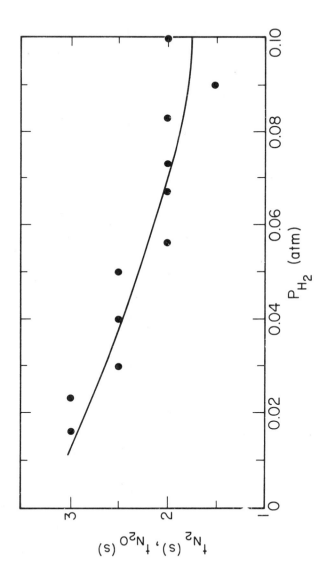

Figure 8. The effect of H_2 partial pressure on the time for the maximum production of N_2 and N_2O, during the reduction of preadsorbed NO: $P_{NO} = 2.8 \times 10^{-3}$ atm, $T = 423$ K, and $t_{NO} = 30$ s.

were performed in which ^{15}NO was substituted for ^{14}NO. Because of problems in resolving products with very nearly identical molecular weights, only the responses for the isotopes of N_2 were followed.

Two experiments were performed. In the first, the catalyst was initially exposed to a $^{15}NO/Ar$ mixture for 60 s after which this mixture was replaced by a $^{14}NO/H_2/Ar$ mixture. The responses for $^{14}N_2$ and $^{14}N^{15}N$ were recorded as functions of time, but the responses for $^{15}N_2$ could not be resolved from that for ^{14}NO, since the two species have nearly the same value of m/e. To establish the $^{15}N_2$ response, the experiment was performed a second time but now reversing the order in which ^{14}NO and ^{15}NO were introduced. The response for the $^{14}N_2$ signal in this case is identical to that for $^{15}N_2$ in the first case. Figure 9 illustrates the partial pressure responses for $^{14}N_2$, $^{14}N^{15}N$, and $^{15}N_2$ observed following introduction of the H_2-containing stream. The $^{15}N_2$ response begins immediately but then passes through a maximum as the adsorbed ^{15}NO is consumed. The $^{14}N^{15}N$ response exhibits an induction period of about 2.5 s, after which it also rises rapidly and then passes through a maximum. The $^{14}N_2$ response begins after a 2 s delay and rises monotonically to a maximum value in about 9 s.

The second experiment was performed to determine the responses of $^{14}N_2$, $^{14}N^{15}N$, and $^{15}N_2$ after a $^{14}NO/H_2/Ar$ mixture was replaced by a $^{15}NO/H_2/Ar$ mixture. This experiment was initiated by exposing the catalyst to a $^{14}NO/Ar$ mixture for 60 s and then replacing this mixture by a $^{14}NO/H_2/Ar$ mixture. After 2 min of reaction, the $^{14}NO/H_2/Ar$ mixture was replaced by a $^{15}NO/H_2/Ar$ mixture, and the responses for $^{14}N_2$ and $^{14}N^{15}N$ were followed as functions of time. For the same reason discussed earlier, the response of $^{15}N_2$ could not be resolved from that for ^{14}NO. To determine the $^{15}N_2$ response, the experiment just described was repeated, but the sequence in which ^{14}NO and ^{15}NO were introduced was reversed. The $^{14}N_2$ response obtained in this case was used to represent the $^{15}N_2$ response for the first case. Figure 10 illustrates the responses for $^{14}N_2$, $^{14}N^{15}N$, and $^{15}N_2$, following the substitution of ^{15}NO for ^{14}NO. The $^{14}N_2$ signal is unaffected during the first second after the isotopic substitution but then declines monotonically to zero. It is significant to observe that the shape of the $^{14}N_2$ response for t ⩾ 2 s and the shapes of the $^{14}N^{15}N$ and $^{15}N_2$ responses are virtually identical to those observed in Fig. 9.

Discussion

Two general features of the reaction mechanism can be deduced from the observations made during the reduction of adsorbed NO by H_2. First, the close coincidence of the N_2

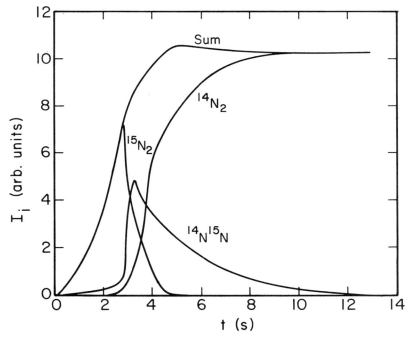

Figure 9.   The gas phase responses for $^{15}N_2$, $^{14}N_2$ and $^{14}N^{15}N$ obtained after substitution of a feedstream containing $^{15}NO/Ar$ by a stream containing $^{14}NO/H_2/Ar$: $P_{NO} = 8 \times 10^{-3}$ atm, $P_{H_2} = 7 \times 10^{-2}$ atm, and $T = 438$ K.

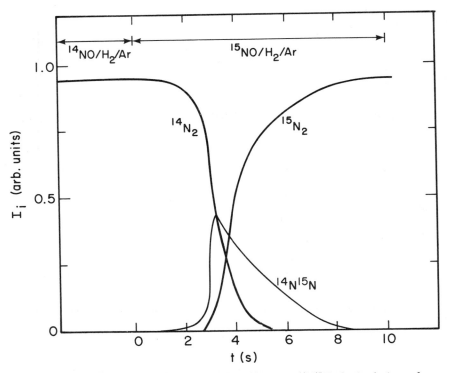

Figure 10.   The gas phase responses for $^{15}N_2$, $^{14}N_2$, and $^{14}N^{15}N$ obtained after sub-stitution of a feedstream containing $^{14}NO/H_2/Ar$ by a feedstream containing $^{15}NO/H_2/Ar$: $P_{NO} = 8 \times 10^{-3}$ atm, $P_{H_2} = 8 \times 10^{-2}$ atm, and $T = 438$ K.

and N_2O peaks suggests that these products share a common rate limiting step, such as, for example, the dissociation of adsorbed NO. Second, the delay in the appearance of the NH_3 and H_2O peaks relative to the N_2 and N_2O peaks suggests that formation of the former pair of products involves adsorbed rather than gaseous hydrogen. This deduction is further supported by the manner in which changes in NO exposure time, H_2 partial pressure, and reaction temperature affect the magnitude of the delay. Figure 11 illustrates a possible mechanism for NO reduction, which is consistent with the qualitative features of the present results and information available in the literature. A discussion of the justification for including specific elementary steps will be presented next.

The associative adsorption of NO is supported by several infrared studies (12-14). Spectra of NO adsorbed on reduced Rh show an intense band at 1660-1740 cm^{-1} and a weaker band at 1830 cm^{-1}. These features are associated with NO adsorbed as $NO_a^{\delta-}$ and NO_a, respectively. If the surface is partially oxidized, a third band is observed at 1910 cm^{-1}, associated with $NO_a^{\delta+}$. During steady-state reduction of NO by H_2, only the $NO^{\delta-}$ form of adsorbed NO is observed (10). The reversibility of NO adsorbtion into this state has been examined by Savatsky and Bell (14). Their work shows that at temperatures above 293 K, equilibrium desorption occurs when a stream of argon is passed over a Rh/SiO_2 catalyst bearing preadsorbed NO.

Evidence for dissociative chemisorption comes from several sources (8,19-21). TPD studies conducted with Rh single crystals (19-21) suggest that a portion of the NO adsorbed at ambient temperatures occurs dissociatively. Further dissociation is presumed to occur at elevated temperatures since N_2 and N_2O are observed during TPD at temperatures slightly above the threshold for NO desorption. A similar behavior has also been observed in TPD studies conducted with Rh/SiO_2 (8). While the exact mechanism of dissociation is not established by these investigations, it seems plausible to propose that dissociation proceeds as indicated by reaction 2 in Fig. 11.

Studies by a number of authors (8,22-27) have shown that H_2 adsorbs dissociatively on Rh and that this process is reversible at the temperatures used in the present studies. As noted earlier, the atomic hydrogen formed by this means is believed to be responsible for the formation of NH_3 and H_2O. Consequently, these products are assumed to be formed by a sequence of Langmuir-Hinshelwood steps. While there is no independent evidence to support this hypothesis for the synthesis of NH_3, recent results reported by Thiel et al. (27) indicate that the formation of H_2O from H_2 and adsorbed O-atoms does proceed via a two step sequence such as that represented by reactions 6 and 7 in Fig. 11.

$$(1) \quad NO + S \rightleftharpoons NO_a$$

$$(2) \quad NO_a + S \longrightarrow N_a + O_a$$

$$(3) \quad H_2 + 2S \rightleftharpoons 2H_a$$

$$(4) \quad NO_a + N_a \longrightarrow N_2O + 2S$$

$$(5) \quad 2N_a \longrightarrow N_2 + 2S$$

$$(6) \quad O_a + H_a \longrightarrow OH_a + S$$

$$(7) \quad OH_a + H_a \longrightarrow H_2O + 2S$$

$$(8) \quad N_a + H_a \longrightarrow NH_a + S$$

$$(9) \quad NH_a + H_a \longrightarrow NH_{2_a} + S$$

$$(10) \quad NH_{2_a} + H_a \longrightarrow NH_3 + 2S$$

Figure 11. The reaction mechanism proposed for the interpretation of transient response experiments.

The specification of reactions 4 and 5 for the formation of
N_2O and N_2 is based on the TPD studies reported by Castner et al.
(19), Campbell and White (20), Baird et al. (21) and Myers and
Bell (8). The work of Myers and Bell has shown specifically that
the formation of N_2O does not occur via a Rideal-Eley process.
It appears more likely that this product is formed via a Langmuir-
Hinshelwood process. To determine whether the mechanism in Fig.
11 could explain the experimental observation reported here, a
model of the reduction process was developed and examined. The
kinetics of NO adsorption and reduction of the adsorbed NO by H_2
can be described by eqns. 1 and 2.

$$\frac{d\overline{P}}{dt} \cdot = \frac{Q}{\varepsilon V_r}(P_i^o - P_i) + \rho_c RTM_m \frac{(1-\varepsilon)}{\varepsilon}\sum_j \nu_{ij}r_j \tag{1}$$

$$\frac{d\overline{\theta}_i}{dt} = \sum_j \nu_{ij}r_j \tag{2}$$

Definitions for the variables and constants appearing in eqns. 1
and 2 are given in the nomenclature section at the end of this
paper. The first of these equations represents a mass balance
around the reactor, assuming that it operates in a differential
manner. The second equation is a species balance written for
the catalyst surface. The rate of elementary reaction j is represented by r_j, and ν_{ij} is the stoichiometric coefficient for component i in reaction j. The relationship of r_j to the reactant
partial pressures and surface species coverages are given by
expressions of the form

$$r_j = k_j \overline{P}_\ell \overline{\theta}_m \tag{3}$$

and

$$r_j = k_j \overline{\theta}_m \overline{\theta}_n \tag{4}$$

for Rideal-Eley and Langmuir-Hinshelwood steps, respectively.
    Equations 1 and 2 can be solved numerically using an
algorithm which handles stiff differential equations (28). Two
sets of boundary conditions are required. For $0 \leqslant t \leqslant t_{NO}$, corresponding to the period during which the reduced catalyst is exposed
to NO, the inlet gas composition is given by

$$P_{H_2}^o = P_{N_2}^o = P_{NO_2}^o = P_{NH_3}^o = P_{H_2O}^o = 0 \tag{5}$$
$$P_{NO}^o = P_{NO}^a$$

and the initial conditions are specified as

$$\overline{P}_{H_2} = 1/2P_{H_2}^r$$

$$\overline{P}_{NO} = 1/2P_{NO}^a$$

$$\overline{P}_{N_2} = \overline{P}_{N_2O} = \overline{P}_{NH_3} = \overline{P}_{H_2O} = 0 \qquad (6)$$

$$\overline{\theta}_H = \sqrt{K_3 P_{H_2}^r}/(1 + \sqrt{K_3 P_{H_2}^r})$$

In eqn. 6, $P_{H_2}^r$ represents the H_2 partial pressure used for the
initial reduction of the catalyst and the subsequent reduction
of adsorbed NO, and P_{NO}^a represents the NO partial pressure
in the feed during NO adsorption. For $t > t_{NO}$, corresponding
to the period of NO reduction, the values of $P_{H_2}^o$ and P_{NO}^o
are changed so that

$$P_{H_2}^o = 1/2 \, P_{H_2}^r$$
$$\qquad\qquad\qquad\qquad (7)$$
$$P_{NO}^o = 0$$

Table I lists the values of the rate coefficients used to
simulate the transient response experiments shown in Figs. 3
through 8. These values were obtained in the following manner
(29). Starting from a set of initial guesses, the values of k
were varied systematically to obtain a fit between the predicted
product responses and those obtained from experiments in which
H_2 was added suddenly to a flow of NO. These experiments while
not described here were identical to that presented in Fig. 9,
with the exception that only ^{14}NO was used. Because of the large
number of parameters in the model, only a rough agreement could
be achieved between experiment and theory even after 500 itera-
tions of the optimization routine (30). The parameter values
obtained at this point were now used to calculate the responses
expected during the reduction of adsorbed NO. These computations
produced responses similar to those observed experimentally
(i.e., Fig. 3) but the appearance of the product peaks in time
did not coincide with those observed. To correct for this, the
values of k_6, k_7, and k_8 were adjusted in an empirical manner.
The criteria used at this point were that the predicted dependences of the delays between the appearance of the NH_3 and H_2O
peaks and the N_2 and N_2O peaks on the duration of NO adsorption
and the H_2 partial pressure during reduction agree as closely
as possible with the dependences found experimentally and shown
in Figs. 5 and 8.

The rate parameters presented in Table I were used together
with the parameter values listed in Table II to predict the
product responses during the adsorption of NO on a hydrogen
covered Rh surface and the subsequent reduction of the adsorbed

Table I.   Rate Coefficients Used to Simulate
the Transient Response Experiments

Step #	k_j
1.	2.8×10^{-4} s^{-1} atm^{-1}
-1.	1.6 s^{-1}
2.	9.3×10^{-1} s^{-1}
3.	2.8×10^{2} s^{-1} atm^{-1}
-3.	1.9×10^{-1} s^{-1}
4.	3.7×10^{-1} s^{-1}
5.	9.9×10^{-1} s^{-1}
6.	6.8 s^{-1}
7.	1.1×10^{1} s^{-1}
8.	6.0 s^{-1}
9.	1.4×10^{3} s^{-1}
10.	1.8×10^{2} s^{-1}

Table II Parameters and Variables Used to
          Simulate the Transient Response
          Experiments

Q = 9.55 cm^3 (STP)/s

ε = 0.96

V_r = 1.6 cm^3

ρ_c = 0.8 gm/cm^3

M_m = 10^{-5} mol Rh sites/gm catalyst

NO. The same parameters were then used to predict the responses
for the isotopic tracer experiments reported in Figs. 9 and 10.
The results of these calculations are presented below.

Figure 12 is representative of the responses predicted for
the product partial pressures both during the period of NO
adsorption and the subsequent reduction of adsorbed NO. Prior to
the addition of NO, the surface is considered to be at equilibrium with the partial pressure of H_2 used to reduce the catalyst.
At time zero the H_2 flow is terminated and NO is added to the Ar
carrier. Reaction products appear immediately due to the
reduction of NO by adsorbed H-atoms. After 30 s of NO exposure,
the flow of NO is terminated and the flow of H_2 is restored. The
introduction of H_2 causes a sudden increase in the product partial
pressures, which then pass through a maximum as the adsorbed NO
is consumed. It is noted that in agreement with experimental
observation (i.e., see Figs. 3 and 4) the calculations predict
that NH_3 is the dominant nitrogen-containing product and that N_2
and N_2O are formed in successively lower concentrations. The
curves in Fig. 12 show further that the production of N_2O and N_2
attain maxima within about 1 to 1.5 s after the introduction of
H_2 but that the production of NH_3 and H_2O occurs about 2 s later.
Both of these features are in good agreement with the behavior
of the experimental data.

The responses for $\bar{\theta}_H$, $\bar{\theta}_{NO}$, and $\bar{\theta}_v$ are shown in Fig. 12. The
value of $\bar{\theta}_H$, which is initially 0.925, decreased during the period
of NO exposure at the same time that the value of $\bar{\theta}_{NO}$ increases
rapidly from zero. It should be noted that the predicted rate of
accumulation of adsorbed NO is qualitatively consistent with the
dynamics of the band appearing at 1680 cm^{-1}, associated with $No_a^{\delta-}$
shown in Fig. 4. It is seen, though, that while the experimental
results exhibit a short induction period followed by a rapid rise
in the absorbance of the 1680 cm^{-1} band to a saturation level, the
predicted curve shows a smooth monotonic increase in $\bar{\theta}_{NO}$.

The vacancy coverage, $\bar{\theta}_v$, which is initially equal to 0.075,
rapidly decreases during the initial period of NO exposure but
then very slowly increases. This behavior can be attributed to
the following factors. The first is that $\bar{\theta}_v$ in equilibrium with
0.10 atm of H_2 is larger than $\bar{\theta}_v$ in equilibrium with 0.0028 atm
of NO. Calculating the equilibrium constants for H_2 and NO adsorption and desorption of these gases, given in Table I, one

concludes that $\bar{\theta}_v^{H_2} = 0.075$ and $\bar{\theta}_v^{NO} = 0.2$. It is this difference
in the strength of adsorption which drives the decrease in θ_v.
Figure 13 shows, though, that during the adsorption of NO, $\bar{\theta}_v$
falls below 0.02 and then gradually climbs back up. This increase is due to formation of a variety of surface intermediates
formed during the initial moments of NO adsorption, which are not
released instantaneously as reaction products. The surface

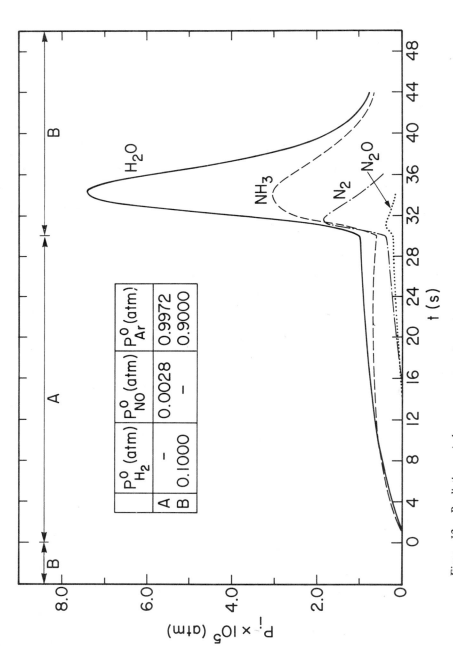

Figure 12. Predictions of the responses for P_{H_2O}, P_{NH_3}, P_{N_2}, and P_{N_2O} during the reduction of preadsorbed NO: $P_{H_2} = 1.0 \times 10^{-1}$ atm, $P_{NO} = 2.8 \times 10^{-3}$ atm, $T = 423$ K, and $t_{NO} = 30$ s.

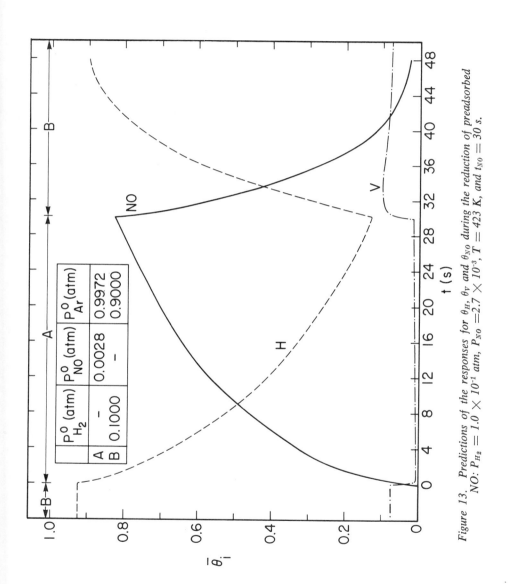

Figure 13.   Predictions of the responses for θ_H, θ_V, and θ_{NO} during the reduction of preadsorbed NO: $P_{H_2} = 1.0 \times 10^{-1}$ atm, $P_{NO} = 2.7 \times 10^{-3}$, $T = 423$ K, and $t_{NO} = 30$ s.

coverage by these species reduces the vacancy fraction. As time
progresses the intermediates are consumed, and the surface
gradually exposes a greater number of vacant sites.

Upon introduction of H_2, following NO adsorption, the magni-
tude of $\overline{\theta}_H$ immediately increases while that of $\overline{\theta}_{NO}$ decreases.
The first of these changes reflects the difference between the
rates of H_2 adsorption and consumption in the reduction of NO.
The decrease in $\overline{\theta}_{NO}$ occurs for two reasons: a) inhibition of the
readsorption of desorbing NO as a consequence of H_2 adsorption;
and b) the consumption of adsorbed NO by reduction. It should be
noted that the dissociation of NO, and hence the rate of NO
reduction is accelerated by the creation of vacant sites. The
increase in θ_v, seen in Fig. 13, can be ascribed to a consumption
of chemisorbed NO and the fact that $\overline{\theta}_v^{H_2}$ is greater than $\overline{\theta}_v^{NO}$,
as was discussed earlier.

Figure 14 shows the computed responses for $\overline{\theta}_N$, $\overline{\theta}_O$, and $\overline{\theta}_{OH}$.
These coverages increase slowly during NO adsorption but rise
rapidly and pass through a maximum when H_2 is added to the flow.
The responses for $\overline{\theta}_{NH}$ and $\overline{\theta}_{NH_2}$ are similar in shape to the
response for θ_{OH} but are significantly smaller in magnitue
and, hence, have not been shown in Fig. 14. It is significant
to note that the maxima in $\overline{\theta}_O$ and $\overline{\theta}_N$ occur at times which
coincide with the time at which the rates of N_2 and N_2O reach
a maximum. Similarly, the maxima in $\overline{\theta}_{OH}$ and $\overline{\theta}_{NH_2}$ appear at times
nearly identical to the times at which the rates of H_2O and NH_3
formation reach a maximum.

Calculations similar to those just discussed were also
carried out for NO exposure times between 5 and 30 s. Figure 15
illustrates the effects of NO exposure time on the predicted
maximum intensity of each product peak. The curves appearing in
this figure may be compared with the experimental results shown
in Fig. 4. It is noted that while the distribution of products
predicted for each NO exposure time is qualitatively consistent
with that observed experimentally, the shape of the product peak
intensity curves is not. The experimental data show a rapid
initial increase which is followed by the attainment of a broad
maximum. By contrast, the predicted curves show a slow monotonic
increase.

The effect of NO exposure time on the time at which the N_2
and N_2O signals attain a maximum is shown in Fig. 16. It is seen
that the model of NO reduction predicts that N_2 formation peaks
about 0.5 s after the peak in the N_2O formation and that the peak
times for both products decline by about 0.5 s as the NO exposure
time is increased from 5 to 30 s. These trends are in good
agreement with the data. It should be noted that since a product
analysis could be taken only once every 0.5 s, it was not possible
to determine product peak positions with an accuracy of better
than 0.5 s. Consequently, both the predicted difference between

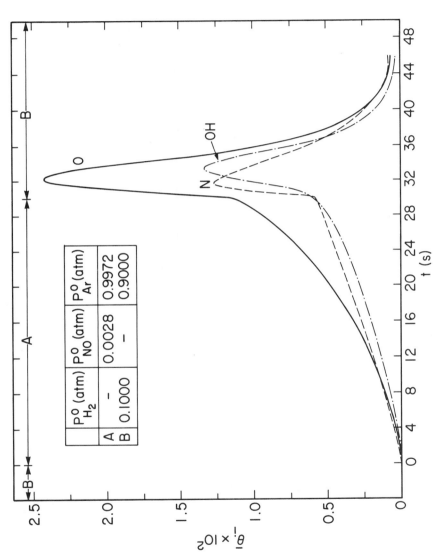

Figure 14. Predictions of the responses for θ_O, θ_N, and θ_{OH} during the reduction of preadsorbed NO: $P_{H_2} = 1.0 \times 10^{-1}$ atm, $P_{NO} = 2.8 \times 10^{-3}$ atm, $T = 423$ K, and $t_{NO} = 30$ s.

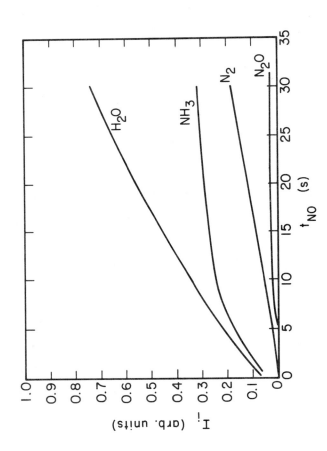

Figure 15. The predicted effects of NO exposure time on the maximum intensities of H_2O, NH_3, N_2, and N_2O, during the reduction of preadsorbed NO: $P_{H_2} = 1.0 \times 10^{-1}$ atm, $P_{NO} = 2.8 \times 10^{-3}$ atm, and $T = 423$ K.

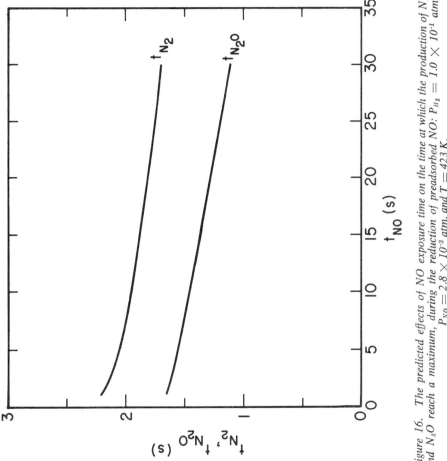

Figure 16. The predicted effects of NO exposure time on the time at which the production of N_2 and N_2O reach a maximum, during the reduction of preadsorbed NO: $P_{H_2} = 1.0 \times 10^{-1}$ atm, $P_{NO} = 2.8 \times 10^{-3}$ atm, and $T = 423$ K.

t_{N_2} and t_{N_2O} and the change in these times with the duration of NO^2 adsorption lie within the limits of uncertainty for the experimental data.

The predicted effects of NO adsorption time on the time delay between the maxima in NH_3 and N_2 production and the maxima in H_2O and N_2 production are shown in Fig. 17. It is seen that the predictions bound the experimental observations and show the the proper trend with increasing NO exposure.

Calculations were also performed to assess the ability of the model to properly predict the effects of H_2 partial pressure on the product peak intensities and the time at which each peak occurs. Figure 18 shows the predicted effect of H_2 partial pressure on the product peak intensities. In contrast to the

trend observed experimentally and illustrated in Fig. 6, the model predicts that intensities decline with increasing parital pressure. Nevertheless, the model does show that the changes are small, and it provides a qualitatively correct picture of the product distribution.

Figure 19 shows the predicted and experimentally observed time at which the formation of N_2 reaches a maximum. As may be seen, the model provides a reasonably good description of the experimentally observed trend. The effect of H_2 partial pressure on the time delay between the maximum in the production of H_2O or NH_3, and the maximum production of N_2 is illustrated in Fig. 20. Here too, the calculated curves are in reasonable agreement with the experimental data and it is noted that the model predicts the same time delay for both NH_3 and H_2O under most circumstances.

To further test the model, calculations were performed to simulate the isotopic tracer experiments presented in Figs. 9 and 10. It should be noted that while the tracer experiments were performed at 438K, the rate coefficients used in the model were chosen to fit the experiments in which chemisorbed NO was reduced at 423 K. Figures 21 and 22 illustrate the nitrogen partial pressure and surface coverage responses predicted for an experiment in which ^{15}NO is substituted for ^{14}NO at the same time that H_2 is added to the NO flow. Similar plots are shown in Figs. 23 and 24 for an experiment in which ^{14}NO is substituted for ^{15}NO during steady-state reduction.

Comparison of Figs. 21 and 23 with Figs. 9 and 10 shows that the predicted partial pressure responses for the N_2 isotopes are in fair agreement with those observed experimentally. The principal differences are that the experimental $^{14}N^{15}N$ response is significantly more asymmetric than the corresponding theoretical response and that all of the experimental transients occur over a shorter time frame than would be deduced from the theoretical results. The strong asymmetry of the experimentally observed $^{14}N^{15}N$ response may be due to the fact that ^{14}N and ^{15}N containing adspecies are not randomly mixed but, rather, exist

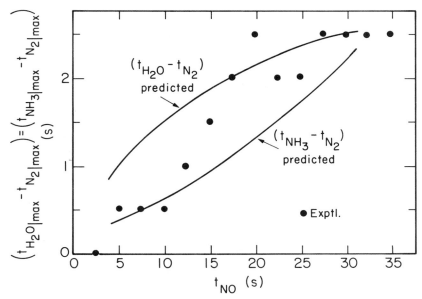

Figure 17.   Comparison of the predicted and the observed effects of NO exposure time on time delay in the maximum production of H_2O and NH_3 relative to N_2, during the reduction of preadsorbed NO: $P_{H_2} = 1.0 \times 10^{-1}$ atm, $P_{NO} = 2.8 \times 10^{-3}$ atm, and $T = 423$ K.

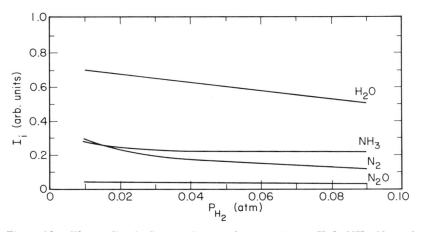

Figure 18.   The predicted effects of P_{H_2} on the intensities of H_2O, NH_3, N_2, and N_2O at maximum production, during the reduction of preadsorbed NO: $P_{NO} = 2.8 \times 10^{-3}$ atm; $T = 423$ K; and $t_{NO} = 30$ s.

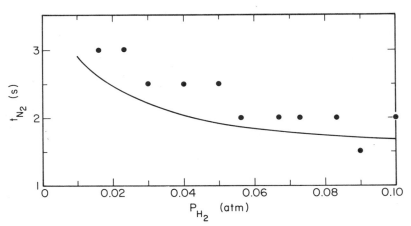

Figure 19.   Comparison of the predicted and the observed effects of P_{H_2} on the time of maximum N_2 production, during the reduction of preadsorbed NO: $P_{NO} = 2.8 \times 10^{-3}$ atm, $T = 423$ K, and $t_{NO} = 30$ s.

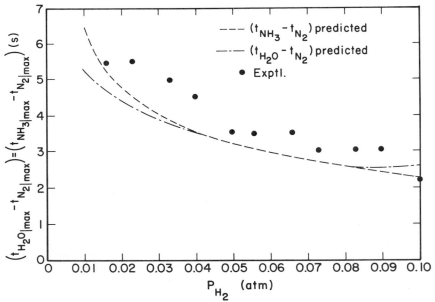

Figure 20.   Comparison of the predicted and the observed effects of P_H on the time delay in the maximum production of H_2O and NH_3 relative to N_2, during the reduction of preadsorbed NO: $P_{NO} = 2.8 \times 10^{-3}$ atm, $T = 423$ K, and $t_{NO} = 30$ s.

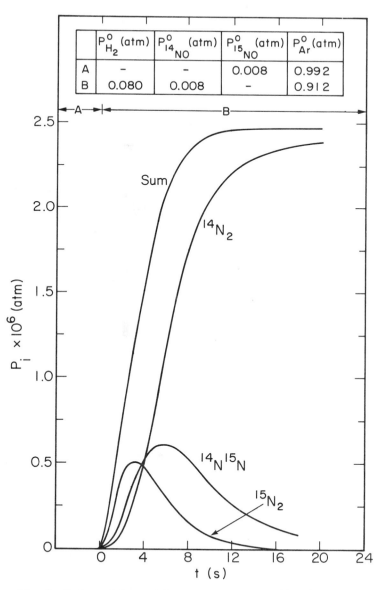

	$P^0_{H_2}$ (atm)	$P^0_{^{14}NO}$ (atm)	$P^0_{^{15}NO}$ (atm)	P^0_{Ar} (atm)
A	–	–	0.008	0.992
B	0.080	0.008	–	0.912

Figure 21.   Predictions of the partial pressure responses for $^{14}N_2$, $^{15}N_2$, and $^{14}N^{15}N$ following substitution of a feedstream containing $^{15}NO/Ar$ by a stream containing $^{14}NO/H_2/Ar$: $P_H = 8.0 \times 10^{-2}$ atm, $P_{NO} = 8.0 \times 10^{-3}$ atm, and $T = 423$ K.

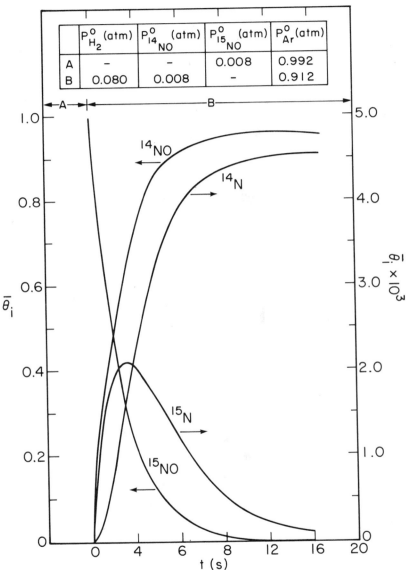

	$P^0_{H_2}$ (atm)	$P^0_{{}^{14}NO}$ (atm)	$P^0_{{}^{15}NO}$ (atm)	P^0_{Ar} (atm)
A	–	–	0.008	0.992
B	0.080	0.008	–	0.912

Figure 22. Predictions of the surface coverage responses for $\bar{\theta}_{{}^{14}NO}$, $\bar{\theta}_{{}^{15}NO}$, $\bar{\theta}_{{}^{14}N}$, and $\bar{\theta}_{{}^{14}N}$ following substitution of a feedstream containing $^{15}NO/Ar$ by a stream containing $^{14}NO/H_2/Ar$: $P_{H_2} = 8.0 \times 10^{-2}$ atm, $P_{NO} = 8.0 \times 10^{-3}$ atm, and $T = 423$ K.

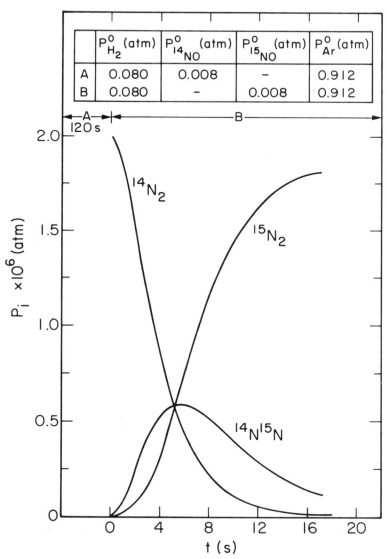

Figure 23.   Predictions of the partial pressure responses for $^{14}N_2$, $^{15}N_2$, and $^{14}N^{15}N$ following substitution of a feedstream containing $^{14}NO/H_2/Ar$ by a feedstream containing $^{15}NO/H_2/Ar$: $P_{H_2} = 8.0 \times 10^{-2}$ atm, $P_{NO} = 8.0 \times 10^{-3}$ atm, and $T = 423$ K.

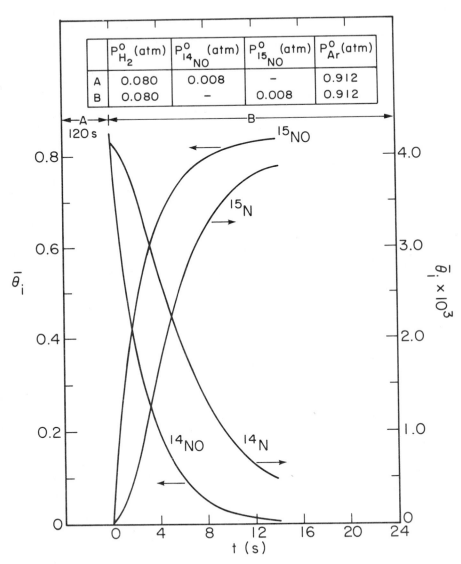

Figure 24.   Predictions of the responses for $\bar{\theta}_{15NO}$, $\bar{\theta}_{14NO}$, $\bar{\theta}_{15N}$, and $\bar{\theta}_{14N}$ following substitution of a feedstream containing $^{14}NO/H_2/Ar$ by a feedstream containing $^{15}NO/H_2/Ar$: $P_{H_2} = 8.0 \times 10^{-2}$ atm, $P_{NO} = 8.0 \times 10^{-3}$ atm, and $T = 423$ K.

in small patches. Such non—uniform mixing would impede the initial formation of the isotopically mixed N_2. The shorter duration of the experimentally observed resonses is not surprising in view of the fact that the experiments were performed at 438 K whereas the calculations were done for a temperature of 423 K.

Conclusions

Transient response experiments have revealed that the formation of N_2 and N_2O during NO reduction by H_2 over Rh proceeds without the intervention of H_2. By contrast, the formation of NH_3 and H_2O involves the reactions of dissociatively chemisorbed H_2 with N and O atoms, respectively. The results obtained from experiments involving the reduction of adsorbed NO and isotopic substitution of ^{15}NO for ^{14}NO can be interpreted on the basis of the reaction mechanism presented in Fig. 11. Key elements of this mechanism are that NO is adsorbed reversibly into a molecular state, that reduction is initiated by the dissociation of molecularly adsorbed NO, and that all products are formed via Langmuir-Hinshelwood process.

Acknowledgment

This work was supported by a grant from the NSF (CPE-7826352).

Nomenclature

k_j —   Rate coefficient for reaction j.
M_m —   Moles of Rh surface sites per gram of catalyst [mol/gm].
P_i^o —   Partial pressure of component i at the reactor inlet [atm].
P_i —   Partial pressure of component i at the reactor outlet [atm].
$\overline{P}_i$ —   Average partial pressure of component i in the reactor, $\overline{P}_i \equiv (P_i^o + P_i)/2$ [atm].
Q —   Total volumetric flow rate [cm^3/s].
r_j —   Rate of reaction j [s^{-1}].
R —   Gas constant [atm $\cdot$ cm^3/K $\cdot$ mol].
t —   Time [s].
T —   Temperature [K].
V_r —   Reactor volume [cm^3].
ε —   Reactor void fraction.
$\overline{\theta}_m$ —   Surface coverage.
v_{ij} —   Stoichiometric coefficient for component i in reaction j.
ρ_c —   Catalyst density [gm/cm^3].

Literature Cited

1. Kobylinski, T. P. and Taylor, B. W., J. Catal. 33, 376 (1974).
2. Klimisch, R. J and Larson, J. G., Eds., "The Catalytic Chemistry of Nitrogen Oxides", Plenum Press, New York, 1975.
3. Shelef, M., Cat. Rev.-Sci. Eng. 11, 1 (1975).
4. Schlatter, J. C. and Taylor, K. C., J. Catal. 49, 42 (1977).
5. Hegedus, L. L., Summers, J. C., Schlatter, J. C., and Baron, K., J. Catal. 56, 321 (1979).
6. Yao, H. C., Yu Yao, Y. F., and Otto, K., J. Catal. 56, 21 (1979).
7. Summers, J. C. and Baron, K., J. Catal. 57, 380 (1979).
8. Myers, E. C. and Bell, A. T., subbmitted to J. Catal.
9. Kobayashi, H. and Kobayashi, M., Catal. Rv.-Sci. Eng., 10, 139 (1974).
10. Bennett, C. O., Catal. Rev.-Sci. Eng. 13, 121 (1976).
11. Tamaru, K., "Dynamic Heterogeneous Catalysis", Academic Press, New York, 1978.
12. Arai, H. and Tominaga, H., J. Catal. 43, 131 (1976).
13. Solymosi, F. and Sarkany, J., Appl. Surface Sci. 3, 68 (1979).
14. Savatsky, B. J. and Bell, A. T., submitted to J. Catal.
15. Connely, N. G., Inorg. Chim. Acta Rev., 47 (1972).
16. Nappier, Jr., T. E., Meek, D. W., Kirchner, R. H. and Ibers, J. A., J. Am. Chem. Soc. 95, 4194 (1973).
17. Goldberg, S. Z., Kubiak, C., Meyer, C. D. and Eisenberg, R., Inorg. Chem. 14, 1650 (1975).
18. Hecker, W. C. and Bell, A. T., unpublished results.
19. Castner, D. G., Sexton, B. A. and Somorjai, G. A., Surface Sci. 71, 519 (1978).
20. Campbell, C. T. and White, J. M., Appl. Surface Sci. 1, 347 (1978).
21. Baird, R. J., Ku, R. C. and Wynblatt, P., Surface Sci. 97, 346 (1980).
22. Mimeault, V. J. and Hansen, R. S., J. Phys. Chem. 45, 2240 (1966).
23. Zakumbaeva, G. D. and Omaskev, Kh. G., Kin. Katal. 18, 450 (1977).
24. Edwards, S. M., Gasser, R. P. H., Green, D. P., Hawkins, D. S. and Stevens, A. J., Surface Sci. 72, 213 (1978).
25. Kawasaki, K., Shibata, M., Miki, H., and Kioka, T., Surface Sci. 81, 370 (1979).
26. Yates, Jr., J. T., Thiel, P. A. and Weinberg, W. H., Surface Sci. 84, 427 (1979).
27. Thiel, P. A. Yates, Jr., J. T., and Weinberg, W H., Surface Sci. 90, 121 (1979).

28.   Hindmarsh, A., "Gear: Ordinary Differential Equation Solver",
      UDID-3001, Rev. 1, August 20, 1972, Computer Center Library,
      University of California, Berkeley, CA.
29.   Savatsky, B. J., Ph.D. Thesis, Department of Chemical
      Engineering, University of California, Berkeley, CA, 1981.
30.   Nelder, J. A. and Mead, R., Computer J. $\underline{7}$, 308 (1964).

RECEIVED August 6, 1981.

The Use of Molybdenum in Automotive Three-Way Catalysts

H. S. GANDHI, H. C. YAO, and H. K. STEPIEN

Ford Motor Company, Dearborn, MI 48121

Abstract

Recent developments in three-way catalyst (TWC) technology may have potential to decrease the amount of rhodium needed in TWC formulations. Current TWC formulations contain platinum (Pt) and rhodium (Rh). The amount of Rh used in the TWCs, that have desired durability, is considerably higher than the mine ratio of Pt/Rh of 17-19. For large scale vehicle application, it is necessary to find ways to minimize the use of this scarce material. Recent findings show that improved net NO_x activity, with minimum NH_3 formation, is accomplished by the incorporation of molybdenum oxide in Pt and Pt-Rh catalysts. Temperature programmed reduction and IR studies were carried out to investigate the interaction of MoO_3 with PtO_2 and with the γ-Al_2O_3 support. The results are used to explain the improved selectivity for the NO to N_2 reaction and the decreased CO poisoning of molybdenum-containing Pt catalysts when compared to pure Pt catalysts.

Introduction

Three-way catalysts (TWC), or equilibrium catalysts are designed for the simultaneous control of three automotive pollutants: NO, CO, and hydrocarbons (HC). Since rhodium has the desired selectivity for the reduction of NO to N_2 with minimum NH_3 formation and is also selective in reducing NO near stoichiometry or under slightly oxidizing conditions[1, 2, 3], it is one of the active metal components used in TWC formulations.[4] The amount of rhodium used in the TWCs having the desired durability is considerably higher than the mine ratio of Pt/Rh of 17-19. Therefore, for large-scale vehicle application it is necessary to find ways to minimize the use of this scarce material. Ruthenium, which has the desired selectivity for the NO to N_2 reaction, has not found use in TWC formulations because of the potential loss of Ru as RuO_3. The attempts to stabilize Ru against volatilization have been partially successful.[5, 6, 7] However, the uncertainty of being able to achieve virtually zero ruthenium emissions, under varying driving conditions, has discouraged the use of Ru in TWC.

This paper describes studies on the use of molybdenum in TWC formulations as a substitute for part of the amount of Rh needed.

Experimental

Catalysts: Catalysts used in the flow reactor study for activity and selectivity measurements are described in Table I. They were monolithic catalysts with a fresh BET area of 15 m^2/g.

Catalysts used in the temperature programmed reduction (TPR) study were granular and are described in Table II. These catalysts were made by agglomerating Dispal-M γ-Al_2O_3 powder (180 m^2/g BET area, Continental Oil Co.) and subsequently wetting, drying and calcining at 600°C for 5 h. The resulting solid mass was crushed and sieved and the 0.5 - 1.0 mm diam. fraction was impregnated with an $(NH_4)_2Mo_2O_7$ solution of desired concentration, dried and calcined at 500°C in air for 2 h to obtain the MoO_3/γ-Al_2O_3 sample. Part of this sample was impregnated again with a H_2PtCl_6 solution of desired concentration, dried and calcined at 500°C in air for 2 h to get the PtO_2-MoO_3/γ-Al_2O_3 sample. For supported Pt catalysts, the granular γ-Al_2O_3 support was impregnated with a H_2PtCl_6 solution, dried and calcined at 500°C in air for 5 h to obtain the PtO_2/γ-Al_2O_3 sample. The pure PtO_2 powder (40 mesh) was a commercial sample obtained from Alpha Chemicals, Ventron Corporation.

Apparatus and Procedure: Flow Reactor for Activity and Selectivity Measurements: The experimental apparatus, testing procedures and methods for analyzing the gas streams were identical to those used in ref. (4). The activity and selectivity of the catalysts were measured with a simulated exhaust gas mixture containing NO, CO, O_2, H_2, C_3H_6, C_3H_8, CO_2, H_2O, and N_2. The CO/H_2 ratio was kept at 3. Propylene and propane were used to represent fast-burning and slow-burning hydrocarbons, respectively. The C_3H_6/C_3H_8 ratio was 2. The synthetic exhaust gas composition and a schematic diagram of the flow-reactor used are also given in ref. (4). In addition, the gas mixture also contained 20 ppm of SO_2.

TABLE I

DESCRIPTION OF MONOLITHIC CATALYSTS[*] USED IN FLOW REACTOR STUDY

Catalyst	Composition wt %			
	Pt	**Pd**	**Rh**	**Mo**
Pt	0.18	-	-	-
MoO_3	-		-	2
$Pt-MoO_3$	0.25		-	2
$Pd-MoO_3$		0.2	-	2
$Pt-Rh-MoO_3$	0.25	-	0.03	2
Pt-Rh	0.12		0.012	-

TABLE II

DESCRIPTION OF THE GRANULAR CATALYST USED IN TPR STUDY

Catalyst	Composition, wt %	
	Pt	**Mo**
$Pt/\gamma-Al_2O_3(A)$	13.8	-
$Pt/\gamma-Al_2O_3(B)$	1.6	-
$Mo/\gamma-Al_2O_3(A)$	-	10.1
$Mo/\gamma-Al_2O_3(B)$	-	3.4
$Pt-Mo/\gamma-Al_2O_3(A)$	1.6	3.4
$Pt-Mo/\gamma-Al_2O_3(B)$	3.2	3.4
$Pt-Mo/\gamma-Al_2O_3(C)$	13.4	3.4

---

[*] Catalysts were prepared on Corning monoliths with 300 square cells/in^2. The fresh BET area of catalyst was 15 m^2/g. ($\gamma-Al_2O_3$= 10 wt % of the uncoated monolith). Active metal composition as wt % of the total monolith weight including washcoal.

The redox ratio R, used to express the oxidizing or reducing characteristics of the exhaust gas mixture, is defined as follows:

$$R = \frac{p_{CO} + p_{H_2} + 3np_{C_nH_{2n}} + (3n + 1)p_{C_nH_{2n+2}}}{p_{NO} + 2p_{O_2}}$$

where p is the partial pressure of the indicated species. Therefore, $R > 1$ represents an overall reducing gas mixture and $R = 1$ represents a stoichiometric gas mixture. The activity and the selectivity measurements were made in a flow reactor with the synthetic gas mixture at 1022°F $(550^\circ$C) and a space-velocity of 60,000 h^{-1}, unless otherwise noted.

Temperature Programmed Reduction (TPR): The apparatus and procedure used in the TPR were reported in a previous paper.[8] A small amount of sample (0.02 to 0.20g) was placed in a Vycor glass cell and a gas mixture of 15% H_2 and 85% Ar was passed over the catalyst at a rate of 25 cc/min. The H_2 uptake during the reduction was measured by a thermoconductivity detector. During the reduction the programmed temperature rise of the sample was controlled at a constant rate of 10°C/min.

Infrared (IR) Study: The apparatus and procedure used in the IR study were reported in a previous paper.[9] The sample wafers were prepared by pressing about 200 mg of the Pt/Mo/γ-Al_2O_3 catalyst (3.2 wt % Pt and 3.4 wt % Mo) in a 2.86 cm diam. metal die and cutting to a size of 6 mm x 20 mm.

Results and Discussion

NO, CO And HC Conversions

The steady-state activity and selectivity characteristics of a monolithic MoO_3 (2 wt % Mo) catalyst are shown in Figure 1A. Percent conversions of CO, NO, HC and ammonia formation are shown as a function of redox ratio, R. Clearly molybdenum oxide itself does not have any activity for NO reduction over the entire range of R values from $R = 0.8$ to $R = 2$. The CO and HC conversions are ~ 15% over this range.

Platinum and palladium catalysts are known to have poor selectivity for the NO to N_2 reaction under reducing conditions. Under these conditions a large fraction of NO is converted to NH_3.[10] Figure 1B shows the characteristic high NH_3 formation over a fresh monolithic 0.18 wt % Pt catalyst. For example, at $R = 1.6$ gross NO conversion is 58%, however, the net NO conversion is ~1%.

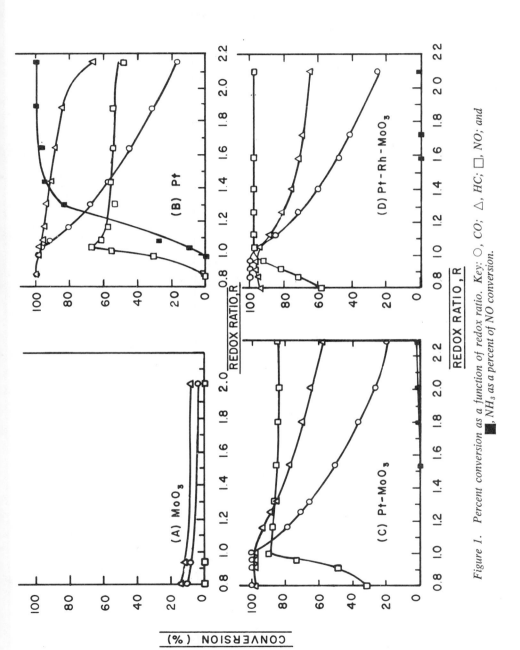

Figure 1. Percent conversion as a function of redox ratio. Key: ○, *CO;* △, *HC;* □, *NO; and* ■, *NH₃ as a percent of NO conversion.*

Figure 1C shows the activity and selectivity characteristics of a monolithic catalyst containing 2 wt% Mo in the form of MoO_3 and 0.25 wt% Pt. The addition of MoO_3 to the Pt catalyst significantly improves the net NO conversion over the entire range of R values. For example, at R = 1.6, 86% net NO conversion is achieved, compared to zero and 1% net NO conversion for MoO_3 only, and Pt only catalysts respectively. The Pt-MoO_3 catalyst also maintains good CO and HC activities.

Similarly the incorporation of 2% Mo in the form of MoO_3 in a 0.2 wt% Pd catalyst significantly improves the net NO_x conversion in the rich region. For example, at R = 1.6, 92% net NO conversion is achieved and the NH_3 formation is only 2% of the converted NO.

The addition of molybdenum oxide to Pt and Pd changes the selectivity characteristic of these catalysts for the NO to N_2 reaction. In the rich region, Pt-MoO_3 and Pd-MoO_3 catalysts suppress NH_3 formation almost completely and thereby yield over 80% net NO_x conversion for R values as high as 1.9 over the Pd-MoO_3 catalyst and as high as 2.3 over the Pt-MoO_3 catalyst. In the lean region (R < 1), the Pt-MoO_3 catalyst is able to reduce NO by CO (and H_2) in the presence of excess O_2. For example, at R = 0.9 the Pt-MoO_3 catalyst converts 50% of the NO compared to no conversion for MoO_3 and ~ 5% conversion for Pt catalysts, respectively. At R = 0.9 the Pd-MoO_3 catalyst exhibits 70% NO conversion. Thus, Pt-MoO_3 and Pd-MoO_3 catalysts minimize NH_3 formation under reducing conditions and improve NO conversion under somewhat oxidizing conditions.

The addition of Rh to Pt-MoO_3 further improves the NO conversion throughout the range of R values. At R = 1 peak NO conversion for the Pt-Rh-MoO_3 catalyst is 98% (Figure 1D) compared to 90% for the Pt-MoO_3 catalyst (Figure 1C).

In addition, the incorporation of MoO_3 substantially decreases NH_3 formation over Pt-Rh catalysts. For example, a catalyst containing 0.19% Pt and 0.017% Rh shows a decrease in NH_3 formation, at R = 1.8, from 20% to ~ 2% by the incorporation of 2% Mo as MoO_3. However, when the amount of Mo is decreased from 2 to 0.3 wt % (i.e. to Mo/Pt atomic ratio of 3) this improvement in the activity or selectivity behavior is lost. This effect suggests the necessity for the Pt and Mo ions to be present in close proximity. By having a large excess of Mo with respect to Pt, the statistical probability for Pt to be located near Mo ions is greatly increased.

Additional experiments were performed to quantify the effect of varying the Mo/Pt ratio on the selectivity for reduction of NO to N_2(or N_2O). The Pt concentration on the monolithic support was maintained at 0.2 wt % and Mo concentration was varied from 0 to 3.4 wt %. The NO-H_2-CO reaction was done over these catalysts at 60,000 h^{-1} space-velocity. The selectivity is defined as:

$$S = \frac{2N_2 + 2N_2O}{2N_2 + 2N_2O + NH_3} \times 100 = \frac{\text{Net NO Conversion}}{\text{Gross NO Conversion}} \times 100$$

The results for NO to N_2 conversion as a function of varying Mo/Pt atomic ratio at 500°C (Figure 2) show the increase of selectivity with Mo concentration, leveling off at Mo/Pt atomic ratio of 15.

Model Reactions Study over Pt and Pt-MoO_3 Catalysts: In order to further understand the activity and selectivity behavior of Mo-containing Pt catalysts, NO-H_2, NO-CO-H_2, and NO-CO model reactions were studied over Pt (0.25 wt %) and Pt-MoO_3 (Pt = 0.25 wt % and Mo = 2 wt %) catalysts.

The study was carried out over a temperature range of 25°C to 600°C at a space-velocity of 60,000 h^{-1}. The relative catalyst activity as expressed by the temperature required for 50% and 80% NO conversion is given in Table III. For both Pt and Pt-MoO_3 catalysts, the presence of CO in the feedgas poisons the Pt sites, as evidenced by an increase in the 50% and 80% NO conversion temperatures. However, the CO poisoning effect is much stronger for pure Pt catalysts than for a mixed Pt-MoO_3 catalyst. For example, for the NO-CO reaction, 50% NO conversion can not be reached over the Pt catalyst, while 345°C and 360°C temperatures are required for 50% and 80% NO conversions, respectively, over the Pt-MoO_3 catalyst.

Table IV shows the striking difference in selectivity behavior between Pt and Pt-MoO_3 catalysts for NO-H_2 and NO-H_2 and NO-H_2-CO reactions. For the NO-H_2-CO reaction at 600°C, only 12% of the reacted NO is converted to N_2 over the Pt catalyst compared to 100% for the Pt-MoO_3 catalyst.

Sulfur Dioxide and Propane Oxidation

Platinum catalysts are known for their good activity for SO_2[11,12] and C_3H_8[13,14] oxidation. Rhodium on the other hand is a poor catalyst for these reactions. Therefore, these reactions were studied on Pt and Pt-

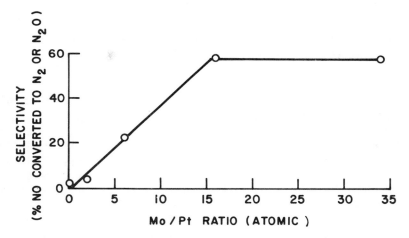

Figure 2.   Selectivity as a function of Mo to Pt ratio
(Pt, 0.2 wt %; R, 1.6) at 500°C.

TABLE III

TEMPERATURE (°C) FOR 50% AND 80% GROSS NO CONVERSION IN NO-H_2, NO-H_2-CO AND NO-CO REACTIONS

Catalysts	NO-H_2 $T_{50\%}$	NO-H_2 $T_{80\%}$	NO-H_2-CO $T_{50\%}$	NO-H_2-CO $T_{80\%}$	NO-CO $T_{50\%}$	NO-CO $T_{80\%}$
Pt(0.25%)	60	100	360	500	NR	NR
Pt-MoO_3 (Pt=0.25%, Mo=2%)	175	185	250	350	345	360
Feedgas	NO = 0.1% CO = 0 H_2 = 1%		NO = 0.1% CO = 0.75% H_2 = 0.25%		NO = 0.1% CO = 1% H_2 = 0	

TABLE IV

SELECTIVITY (%) FOR NO TO N_2(or N_2O) REACTION IN NO-H_2 NO-H_2-CO REACTIONS AS A FUNCTION OF TEMPERATURE °C

T°C	Pt NO-H_2	Pt NO-H_2-CO	Pt-MoO_3 NO-H_2	Pt-MoO_3 NO-H_2-CO
250	0	0	0	30
300	0	0	0	43
350	0	0	0	53
400	0	0	12	60
450	0	0	22	67
500	0	2	30	73
550	0	8	38	80
600	0	12	46	100

MoO_3 catalysts to determine whether the Pt-MoO_3 catalyst exhibits Pt-like or Rh-like behavior. The results for SO_2 oxidation over supported Pt and Pt-MoO_3 catalysts are shown in Table V. The Rh-like behavior, i.e., poor SO_2 oxidation activity, is evident. For example, at 500°C and 60,000 h^{-1} space-velocity, SO_2 conversion is 45% for the Pt catalyst compared to 11% for the Pt-MoO_3 catalyst. Table VI shows similarly the extreme difference in propane oxidation activity for these catalysts.

Effect of Oxygen Pulse on Steam-Reforming Activity

The effect of an oxygen pulse on the steam-reforming activity of fresh Pt, Rh, Pt-Rh, and Pt-MoO_3 catalysts was studied both in the presence and absence of 20 ppm SO_2 in the feed-gas (Table VII). The good steam-reforming activity of Pt, Rh, and Pt-Rh catalysts is severely reduced by the addition of 20 ppm of SO_2 in the feed-gas. The incorporation of MoO_3 significantly suppresses the steam-reforming activity of Pt catalysts even in the absence of the SO_2 in the feed-gas. In this case, the Pt-MoO_3 catalyst performance is similar neither to Pt nor Rh catalysts. Upon adding a 30 sec 1% O_2 pulse, the HC conversion improves for all the catalysts. The average values of percent HC converted during five consecutive 30 sec O_2 pulses, both in the presence and absence of SO_2 in the feed-gas, also are given in Table VII. The improved HC conversion lasts for several minutes after the 30 sec O_2 pulse is terminated. The HC conversion values at t = 10, 15, 20 and 30 sec, after the O_2 pulse was ended, are shown in Table VIII. A comparison of HC conversion values shows that the Pt-MoO_3 catalyst performance is somewhat inferior to that of the Pt catalyst under these cyclic conditions.

These results show that incorporation of MoO_3 into a Pt catalyst changes the basic characteristics of the Pt catalyst and suggest the possibility of an interaction between Pt and Mo oxides on the surface of the support. This possibility was investigated by a TPR study with a Pt-MoO_3 catalyst.

Temperature Programmed Reduction (TPR)

The rates of H_2 uptake (in arbitrary units) as a function of temperature for the γ-Al_2O_3 supported PtO_2, MoO_3, and PtO_2-MoO_3 mixtures of various concentrations are plotted in Fig. 3. TPR of pure PtO_2 and MoO_3 supported on γ-Al_2O_3 have been reported.[15,16] Curve A is the TPR of the PtO_2 sample of high concentration (13.8 wt % Pt) in which two

TABLE V

SO_2 OXIDATION OVER Pt AND Pt-MoO_3 CATALYSTS

	% SO_2 Conversion[a]		
Catalyst	400°C	500°C	600°C
Pt (0.2 wt %)	10	45	59
Pt-MoO_3 (Pt = 0.2 wt %) (Mo = 2 wt %)	3	11	16

[a] Feedgas: SO_2 = 0.02%
O_2 = 1%
S.V. = 60,000 h^{-1}

TABLE VI

PROPANE OXIDATION OVER Pt AND Pt-MoO_3 CATALYSTS

Catalyst	$T_{25\%}$	$T_{50\%}$	$T_{75\%}$
Pt (0.2%)	230	267	298
Pt-MoO_3 (Pt = 0.2%, Mo = 2%)	450	640	N.R.

Feedgas: C_3H_8 = 0.05%
O_2 = 2.25%
SO_2 = 20 ppm
N_2 = Balance
S. V. = 60,000 h^{-1}

TABLE VII

EFFECT OF OXYGEN PULSE ON STEAM REFORMING ACTIVITY

		% HC Conversion[A]			
		Steady-State		Oxygen Pulse[B]	
Catalyst ppm SO_2:		0	20	0	20
Pt (0.2%)		70	20	90	73
Rh (0.02%)		80	8	93	72
Pt (0.2%), Rh (0.02%)		89	3	98	84
Pt (0.2%), Mo (2%)		12	4	57	57

[A] Feedgas: C_3H_6 = 0.036%
$\qquad\qquad$ C_3H_8 = 0.018% H_2O = 8%
$\qquad\qquad$ H_2 = 0.5% SO_2 = 0 or 0.002%
$\qquad\qquad$ CO_2 = 8% N_2 = Balance
$\quad$ T = 600°C,   S.V. = 60,000 h^{-1}

[B] Average % HC converted during 5 consecutive
30 sec 1% O_2 pulses.

rate maxima are observed. The one at 10°C has been assigned to the reduction of PtO_2 in the particulate phase and the one near 90°C to the reduction of PtO_2 in the saturated dispersed phase.[15] Curve B is TPR of the PtO_2 sample of low concentration (1.6 wt % Pt or 0.48 μmol/m^2 BET), which is far below the saturation concentration of the dispersed phase (2.2 μmol/m^2 BET). The rate maximum in Curve B has shifted to a higher temperature, ~ 250°C, indicating a stronger PtO_2-γ-Al_2O_3 interaction in this sample. Curve C is the TPR of the MoO_3 sample of high concentration (10.1 wt % Mo). The surface concentration of MoO_3 in this sample is 8.2 μmol/m^2 (BET), of which about 50% is present in the particulate phase.[16] The TPR of this sample shows three rate maxima, indicating the difference in the ease of removal for the different oxygen atoms in MoO_3 by H_2. The first rate maximum at 450°C has been assigned[16] to the removal of only

TABLE VIII

STEAM REFORMING WITH EFFECT OF OXYGEN PULSE (T = 600°C, S.V. = 60,000 h^{-1})

Catalyst	Feedgas HC (ppm)	HC Conv. %	10 sec after (first)	15 sec 30 sec O$_2$ pulse	20 sec O$_2$ pulse (O$_2$ = 1%)	30 sec
			30 sec O$_2$ pulse HC Conv.(%)			
W/O SO$_2$						
0.2% Pt	1500	70	-	-	-	-
0.02% Rh	1600	81	95	94	93	92
0.2% Pt + 0.02% Rh	1500	89	98	97	97	97
0.2% Pt + 2% Mo	1620	12	85	44	17	9
With 20 ppm SO$_2$						
0.2% Pt	1500	21	77	68	61	54
0.02% Rh	1600	8	86	78	69	56
0.2% Pt + 0.02% Rh	1500	3	82	70	62	55
0.2% Pt + 2% Mo	1620	4	72	61	43	32

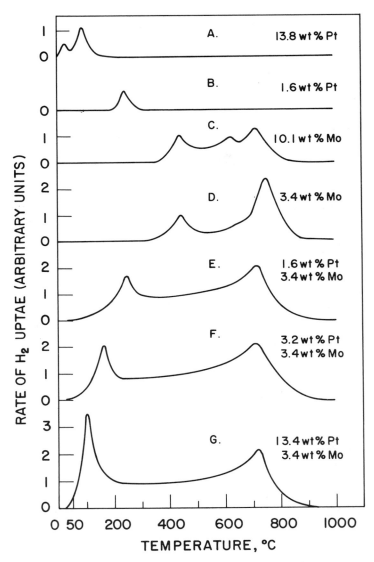

Figure 3.   Rate of H_2 uptake as a function of temperature in the TPR of PtO_2,
MoO_3, and PtO_2–MoO_3 supported on γ-Al_2O_3.

surface capping oxygen atoms. The other two rate maxima at 580°C and 720°C have been assigned to the removal of the other oxygen atoms in the sublayer and in the bulk of MoO_3 in the particulate phase.

In $MoO_3/\gamma-Al_2O_3$ of low concentration (3.4 wt % Mo, curve D), all of the MoO_3 is present in the dispersed phase.[16] The TPR of this sample gives only two rate maxima. The first one at 450°C is the same as that in curve C and has been assigned to the removal of the surface capping oxygen.[16] The second rate maximum at higher temperature (750°C) has been assigned to the oxygen atoms other than the capping oxygen atoms in the dispersed (Figure 3D) phase of MoO_3 on $\gamma-Al_2O_3$.

The presence of PtO_2 in the $Pt-Mo/\gamma-Al_2O_3$ catalyst lowers the reduction temperature of the first rate maximum from 450°C down to 250°C, 160°C or 100°C depending upon the concentration of PtO_2 (curves E-G); however, it does not affect significantly the second rate maximum at about 750°C. Also, note that the rate maximum for the reduction of PtO_2 coincides with the first rate maximum. These results suggest that the interaction between the PtO_2 and the dispersed MoO_3 on $\gamma-Al_2O_3$ is responsible for lowering the temperature for the removal of the capping oxygen atoms and rule out the mechanism of hydrogen spillover contributing to the observed effect. If hydrogen spillover predominates, then it should also lower the temperature for the removal of other oxygen atoms in the dispersed phase of MoO_3. In fact, the interaction lowers only the temperature for the removal of surface capping oxygen indicating that the interaction of PtO_2 involves only the capping oxygen of the dispersed MoO_3. After this reduction the Pt atoms may form a surface complex with the partially reduced Mo ions. For the $Pt-Mo/\gamma-Al_2O_3$ catalysts which are used at temperatures < 600°C, the active complex may be the dispersed $PtMoO_x$ (where X = 2-3), under the reducing conditions. For temperatures > 600°C, the catalyst can have the surface complex of $PtMoO_{2-x}$ where x ranges from 0 to 2 depending upon the temperature of reduction.

Infrared (IR) Study

IR spectra of CO chemisorbed on a $Pt-Mo/\gamma-Al_2O_3$ sample (3.2 wt % Pt, 3.4 wt % Mo) are presented in Figure 4. Spectrum A was taken after the $Pt-Mo/\gamma-Al_2O_3$ sample was degassed at 300°C and followed by CO chemisorption at 25°C. The weaker band at 2180 cm^{-1} has been assigned

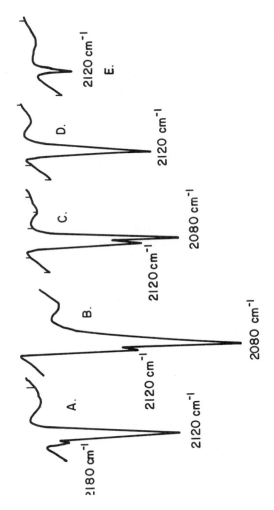

Figure 4.   IR data of CO chemisorbed on a Pt–Mo/γ-Al₂O₃ catalyst.

to CO on Pt ions (Pt oxide) and the stronger band at 2120 cm^{-1} to chemisorbed CO on Pt.[17,18] The sample was subsequently heated in 60 torr CO at 300°C for 30 min and cooled to 25°C, giving spectrum B in Figure 4. In spectrum B, the band at 2180 cm^{-1} has disappeared, the band at 2120 cm^{-1} has weakened, and a strong band at 2080 cm^{-1} has emerged. This strong new band was also found in the spectrum (not shown) taken after CO chemisorption on a reduced (500°C, H_2) $MoO_3/\gamma-Al_2O_3$ sample and was assigned to the CO chemisorbed on reduced Mo ions. Apparently at 300°C, PtO_2 is reduced to Pt^o, and surface MoO_3 is reduced primarily to MoO_2.[19] The CO chemisorbed on the reduced molybdenum oxide is easier to remove when degassed at 200°C and is more reactive with O_2 at elevated temperatures. By exposing this sample to air at 25°C, the band at 2080 cm^{-1} decreases, whereas the band at 2120 cm^{-1} increases slightly (spectrum C). After heating the sample in air at 100°C for 20 min, the band at 2080 cm^{-1} disappears completely while the band at 2120 cm^{-1} further increases (spectrum D). Only after heating in air at 300°C, will the band at 2120 cm^{-1} gradually disappear (spectrum E). At 350°C, the band at 2120 cm^{-1} disappears completely. These results show that CO chemisorbed on Mo ions is more reactive and can be oxidized by air at 25° - 100°C while the CO chemisorbed on Pt is held strongly and can only be oxidized after it is heated near 300°C.

Conclusion

Previously, Ermakov et al.[20,21] studied Pt-Mo/SiO$_2$ catalysts and found that the presence of Mo in Pt-Mo/SiO$_2$ catalysts increased the rate of hydrogenolysis by 4 orders of magnitude compared to Pt/SiO$_2$ catalysts and decreased the electron density of Pt as observed from x-ray electron spectroscopy. They interpreted these data in terms of chemical bonding between Pt and Mo and suggested a Pt^o-Mo^{2+} complex on silica which has catalytic activity and selectivity similar to Rh or Ir rather than to Pt.

Our activity and selectivity results are in good agreement with their findings. Our model reactions study showed that the presence of Mo in a

Pt-Mo/γ-Al$_1$O$_3$ catalyst (1) lowers the activity but increases the selectivity of the Pt catalyst for the NO reduction by H$_2$, and (2) increases the activity for the NO reduction by CO and the activity and selectivity for the NO reduction by CO + H$_2$. Based on the TPR and IR data, we attribute these results to a strong interaction between Pt and Mo oxides on the γ-Al$_2$O$_3$ support. Such interaction facilitates the removal of the surface oxygen atom of the dispersed MoO$_3$ to form a surface Pt^o-Mo^{4+} complex which has activity and selectivity characteristics similar to that of Rh/γ-Al$_2$O$_3$ or Ir/γ-Al$_2$O$_3$ rather than Pt/γ-Al$_2$O$_3$ in the reduction of NO by H$_2$. Both Pt^o and Mo^{4+} sites in the Pt^o-Mo^{4+} complex chemisorb CO, but CO molecules chemisorbed on Mo^{4+} sites are more reactive than CO on the Pt^o sites. Also, CO chemisorbed on Mo^{4+} decreases the affinity of CO for Pt^o sites probably due to the dipole-dipole repulsion between the adsorbed CO molecules on both Pt and Mo^{4+} sites in the Pt^o-Mo^{4+} complex. Both of these effects enhance the activity of the catalyst for NO reduction by CO and the activity and the selectivity by H$_2$ + CO.

Practical Considerations

While the ability of Pt-MoO$_3$ or Pd-MoO$_3$ catalysts to suppress NH$_3$ formation must be preserved, stabilization of MoO$_3$ in these catalysts from volatilization at higher temperatures is necessary under cyclic conditions around a stoichiometric A/F ratio. MoO$_3$ volatilizes at temperatures - 600°C under oxidizing conditions, as can be inferred from the vapor pressure values for MoO$_3$ given in Table IX.[22] Although molybdenum oxides are not considered toxic[23] compared to the noble metal oxides (e.g., RuO$_4$ and OsO$_4$), it is nevertheless desirable to minimize loss of molybdenum by volatilization while maintaining an optimum level of activity and selectivity.

Stabilization of MoO$_3$ is achieved by incorporation of alkaline-earth metal oxides (e.g. BaO, MgO) or rare-earth metal oxides (e.g., La$_2$O$_3$) or by reaction with base metal oxides (e.g., NiO, CoO, CuO etc.). Since MoO$_3$ is an acidic oxide, its reaction with a basic oxide should lead to a mixed oxide

TABLE IX

VAPOR PRESSURE OF MoO_3

T^oC	V.P. (torr)
627	3.9×10^{-3}
727	0.13
827	1.9
929	11.7
1027	53.2

which should be more stable than MoO_3 against volatilization. Interaction between MoO_3 in the dispersed phase[16] and the γ-Al_2O_3 support also lower the lovatility of MoO_3. Work is currently in progress to optimize the incorporation of MoO_3 into TWCs without sacrificing the desired activity and selectivity.

Acknowledgements

We would like to thank Dr. J. T. Kummer for providing helpful consultation throughout this work and Dr. K. M. Adams for obtaining SO_2 and C_3H_8 oxidation data.

Literature Cited

1. Kobylinski, T. P. and Taylor, B. W., J. Catal. 33, 376 (1974).

2. Ashmead, D. R., Campbell, J. S., Davies, P. and Farmery, K., SAE (Soc. Automotive Eng.) Paper 74039 (1974).

3. Taylor, K. C. in "The Catalytic Chemistry of Nitrogen Oxides," R. L. Klimisch and J. G. Larson (Eds.), Plenum Press, New York, 1975, p. 173.

4. Gandhi, H. S., Piken, A. G., Shelef, M. and Delosh, R. G., SAE Transactions, 85, Paper 760201, 901 (1976).

5. Shelef, M. and Gandhi, H. S., Platinum Metals Review 18(1), 1 (1974).

6. Gandhi, H. S., Stepien, H. K. and Shelef, M., SAE Paper 750157 (1975).

7. Gandhi, H. S., Stepien, H. K. and Shelef, M., Material Research Bulletin, 10, 837 (1975).

8. Yao, H. C., Japar, S. and Shelef, M., J. Catal. 50, 407 (1977).

9. Yao, H. C. and Rothschild, W. G., J. Chem. Phys. 68, 4775 (1978).

10. Shelef, M. and Gandhi, H. S., Ind. Eng. Chem., Prod. Res. Develop. 13, 80 (1974).

11. Gandhi, H. S., Yao, H. C., Stepien, H. K. and Shelef, M., SAE Paper 780606 (1978).

12. Williamsom, W. B., Truex, T. J., Gandhi, H. S. Ku, R. C. and Wynblatt, P., AIChE Symposium Series No. 201, 76, 212 (1980).

13. Williamson, W. B., Stepien, H. K. and Gandhi, H. S.; Environ. Sci. & Tech. 14, 319, 1980.

14. Cooper, B. J., Harrison, B., Shutt, E. and Lichtenstein, I., SAE Paper 770367 (1977).

15. Yao, H. C., Sieg, M. and Plummer, H. K., Jr., J. Catal. 59, 365 (1979).

16. Yao, H. C., J. Catal. (in Press).

17. Heyne, H. and Tompkins, F. C., Trans. Faraday Soc. 63, 1274 (1967).

18. Summers, J. C. and Ausen, S. A., J. Catal. 58, 131 (1979).

19. Hall, W. K. and Lo Jacono, M., Proceedings of the Sixth Int. Cong. on Catal. 1, 246 (1977).

20. Ermakov, Yu. I., Ioffe, M . S., Ryndin, Yu. A. and Kuznetsov, B. N., Kinetika i kataliz 16, 807 (!975).

21. Ermakov, Yu. I., Kuznetsov, B. N., Rydin, Yu. A. and Kuplyakin, V. K., Kinetika i Kataliz 15, 1093 (1974).

22. Thermodynamic Properties of Molybdenum Compounds, Bulletin CDB-2 Climax Molybdenum Company, Sept., 1954.

23. Sax, N. I., "Dangerous Properties of Industrial Materials," Van Nostrand Reinhold, p. 939, New York, 1975.

RECEIVED June 26, 1981.

Decomposition of Nitrous Oxide on Magnesium Oxide Compared to That on Cupric Oxide

H. KOBAYASHI and K. HARA

Hokkaido University, Department of Chemical Process Engineering, Sapporo 060, Japan

The decomposition of nitrous oxide over various metal oxides has been widely investigated by many investigators (1-3). Dell, Stone and Tiley (4) have compared the reactivity of metal oxides and shown that in general p-type oxides were the best catalysts and n-type the worst, with insulators occupying an intermediate position. It has been generally accepted (5) that this correlation indicates that the electronic structure of the catalyst is an important factor in the mechanism of the decomposition of nitrous oxide over metal oxides catalysts. The reaction is usually written (4) as

$$N_2O + e^- \text{ (from catalyst)} \rightarrow N_2 + O^- \text{ (ads)} \tag{1}$$

$$2 \; O^- \text{(ads)} \rightleftarrows O_2 + 2 \; e^- \text{ (to catalyst)} \tag{2}$$

$$\text{or} \quad O^- \text{ (ads)} + N_2O \rightarrow N_2 + O_2 + e^- \text{ (to catalyst)} \tag{3}$$

More recently Vijh (6) has shown that the existing data on the catalytic decomposition of nitrous oxide can be fit to an interpretation very similar to that involved in the Sabatier-Balandin concept. The fundamental idea behind Sabatier-Balandin approach is that the reactants get adsorbed on the catalyst to form a surface compound and the energy of the bond between the catalyst and the reactant is related to the catalytic activity in a volcanic manner; i.e., there is a maximum in activity with increasing bond energy. Based on this interpretation, Vijh has suggested that in the ascending branch of the volcano, the activity decreases with increasing bond energy, indicating that the rate-determining step in the N_2O decomposition on these oxides such as CuO, NiO and CoO probably involves rupture of a bond between the catalyst and the reactant. On the other hand, in the descending branch of the volcano, the activity increases with increasing bond energy and this indicates a rate-determining step involving the formation of a bond between the catalyst and the reactant.

0097-6156/82/0178-0163$05.00/0

Since the insulator oxides such as MgO, CaO and Al_2O_3 are involved in this branch, it would be worth examining whether the mechanism of N_2O decomposition on these insulator oxides is different from that on p-type oxides.

In the present paper, the reactions on CuO and MgO are examined and compared.

Experimental

Catalysts. Cupric oxide was prepared by thermal decomposition of reagent grade copper nitrate (Wako Pure Chem.Inc.Ltd.) at 400°C in air for 4 hrs. Magnesium oxide was commercially available reagent grade powder (Kanto Chemical Co.Ltd.). The oxides powders were pressed into tablets and crushed and 24-42 mesh granules were used as catalysts.

Apparatus and Procedure. The kinetic studies of the catalysts were carried out by means of the transient response method (7) and the apparatus and the procedure were the same as had been used previously (8). A flow system was employed in all the experiments and the total flow rate of the gas stream was always kept constant at 160 ml STP/min. In applying the transient response method, the concentration of a component in the inlet gas stream was changed stepwise by using helium as a balancing gas. A Pyrex glass tube microreactor having 5 mm i.d. was used in a differential mode, i.e. in no case the conversion of N_2O exceeded 7 %. The reactor was immersed in a fluidized bed of sand and the reaction temperature was controlled within ± 1°C.

The composition of the gas stream before and after contact with the catalyst was monitored by subjecting aliquots to gas chromatographic analysis using a Ohkura Model 701 gas chromatograph and two columns. For N_2 and O_2, a 1 meter molecular sieve 5A column operating at 65°C was employed ; for N_2O, a 2 meter Porapak Q column operating at 89°C was used.

Reaction. The effect of external diffusion on the kinetics was examined at same W/F over different bed size and was found to be negligible at the gas velocity and temperatures used in the kinetic experiments. By varying the gas velocity of a constant composition gas mixture through a bed, no appreciable change in percentage of N_2O decomposition was observed. The effect of internal diffusion was investigated by measuring the rate of reaction under indentical feed and flow rate conditions for beds of (a) 28-36 mesh and (b) 20-28 mesh catalyst granules. The reaction rates on both catalyst beds were essentially the same, showing evidence of the absence of pore diffusion control. The reactor wall was found to be catalytically inactive at temperatures employed.

Nitrous oxide (N_2O 99.0 %) and helium (He 99.999 %) from commerical cylinders were purified through a dry ice-methanol trap to remove water vapor.

Results

The response of the component B in the effluent stream to a step change in the concentration of A in the feed gas stream is designated as A-B response. When A is increased from nil, A(inc., 0)-B, and decreased to nil, A(dec.,0)-B.

CuO Catalyst. 70.9 g of CuO catalyst was used in the experiments. The temperature at which the decomposition of N_2O became appreciable at steady state was 230°C and the reaction experiments were carried out at 256°C. In Figure 1 is shown a response of N_2, O_2 and N_2O in the outlet flow stream to the step change of N_2O in the inlet gas stream, i.e. $N_2O(dec.,0)-N_2$, 0, N_2O. The response of O_2 and N_2O was found to tail considerably, indicating that these species were adsorbed on the catalyst surface during the reaction at steady state. On the contrary, the N_2 response is quite rapid and suddenly dropped down to nil. It is worth noting that nitrogen did not evolve in spite of the presence of adsorbed nitrous oxide. It would be reasonable to suggest from this findings, therefore, that the decomposition of nitrous oxide is taking place through the immediate reaction of gaseous nitrous oxide impinging on the active site on the surface.

The slow desorption of oxygen will also suggest that the desorption of oxygen from the catalyst could be the slowest step during the steady state reaction.

After the catalyst had been kept in a helium stream for a surfficient time to remove oxygen, the feed was switched over to a stream of $N_2O(17\%)$-He mixture and the $N_2O(inc.,0)-N_2,O_2$ response was followed. The results were shown in Figure 2. Upon changing the stream, the N_2O-N_2 response showed an instantaneous maximum which was followed by a steep decrease showing an overshoot-type response. The N_2O-O_2 response, on the other hand, showed a monotonic increase to attain a new steay state value. This delayed increase of oxygen concentration in the effluent stream indicates that the oxygen produced by the decomposition of nitrous oxide is accumulatively adsorbed on the catalyst surface until a new steady state is attained. These results clearly indicate that the oxygen adsorbed on the surface inhibits the reaction, and that N_2O can decompose only on sites which are free from adsorbed oxygen. In support of this, the effect of oxygen on the rate of N_2O decomposition was examined and the results were shown in Figure 3. The presence of oxygen in the gas phase markedly inhibited the reaction substantiating that the sites which are active for the N_2O decomposition reaction are also active for the adsorption of oxygen. The instantaneous maximum in the N_2O-N_2 response curve, therefore, can be attributed to the high concentration of vacant sites in the initial steady state in the helium stream, and the steep decrease which follows will be caused by blocking of active sites with accumulatively adsorbed oxygen.

Since gaseous nitrous oxide decomposes directly on the active

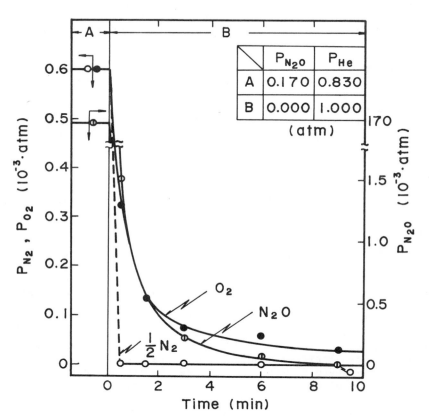

Figure 1. $N_2O(dec. O)-N_2, O_2, N_2O$ response on CuO at 256°C.

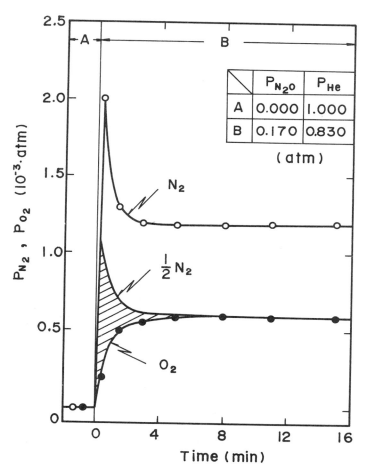

Figure 2. $N_2O(inc.~O)–N_2, O_2$ response on CuO at 256°C.

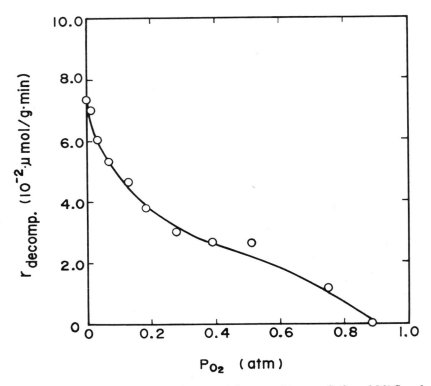

Figure 3. *Effect of oxygen on the rate of decomposition on CuO at 256°C and P_{N_2O} 0.113.*

sites and the nitrogen produced is not adsorbed on the surface, the N_2O-N_2 response curve will represent the decomposition rate of nitrous oxide at each instance. Therefore the graphical integration of the N_2O-N_2 response curve up to the new steady sate gives the total amount of nitrogen produced by the reaction, and hence the amount of oxygen produced by the reaction can be easily estimated by a stoichimetric relationship. The analogous graphical integration of N_2O-O_2 response curve, on the other hand, will give the amount of oxygen desorbed from the catalyst surface. The difference of those amounts of oxygen gives the amount of retained oxygen on the surface in excess to that at initial steady state, and this procedure is visualized in Figure 2. The amounts of adsorbed oxygen at steady states thus estimated in the stream of various partial pressure of nitrous oxide, as designated as $q^{re}(O_2)$, were plotted against corresponding partial pressure of nitrous oxide in the inlet stream. The results are shown in Figure 4 together with the amounts of adsorbed N_2O at steady states which were estimated by the graphical integration of N_2O (dec.,$0)-N_2O$ curves. The measurements of isothermal adsorption of oxygen on the same catalyst were accomplished in separate experiments and the equilibrium amounts of adsorbed oxygen, $q^{eq}(2)$, are also plotted in the same figure. The corresponding oxygen pressure on the abscissa were so chosen that the oxygen pressure is equal to the average partial pressure of oxygen across the catalyst bed curing the reaction with the inlet concentration of the corresponding partial pressure of nitrous oxide on the abscissa. Since the adsorption of gaseous oxygen on this catalyst was revealed to be dissociative at 256°C and obeyed the Langmuir isotherm, the maximum amount of adsorption can be easily estimated. Based on this value, the fraction of vacant site, θ_V, was also estimated at various P_{N_2O} and plotted in the same figure.

It can be seen from the figure that the $q^{re}(O_2)$ is much higher than $q^{eq}(O_2)$, which strongly suggests that the desorption of oxygen from the catalyst surface is the slowest step in this reaction.

According to the results obtained above, the following mechanism seems to be reasonable for this reaction. This mechanism, in which oxygen desorption is the slowest step, has been suggested by many investigators.

$$N_2O + e^- \text{ (from catalyst)} \rightarrow N_2 + O^- \text{ (ads)} \tag{1}$$

$$2 \ O^- \text{ (ads)} \rightleftarrows O_2 + 2 \ e^- \text{ (to catalyst)} \tag{2}$$

As has been noticed above nitrogen was not produced unless N_2O existed in the gas phase even though the adsorbed N_2O existed on the surface. This fact may imply the possibility of the following steps which have been suggested by several investigators.

$$N_2O + O^- \text{ (ads)} \rightarrow N_2 + O_2 + e^- \text{ (to catalyst)} \tag{3}$$

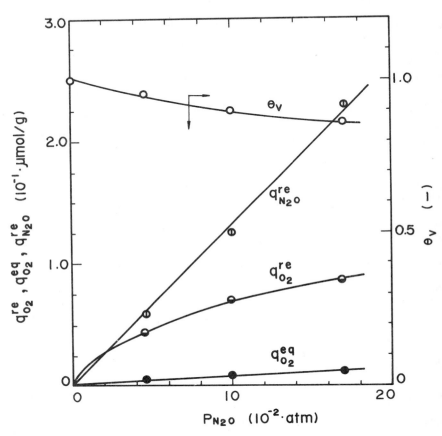

Figure 4.   Amounts of adsorbed O_2 $q^{re}(O_2)$ and N_2O $q^{re}(N_2)$ during the reaction and equilibrium amount of adsorbed O_2 $q^{eq}(O_2)$ obtained by O_2 adsorption at $256°C$. θ_V is the vacant fraction of active sites during the reaction at steady state on CuO.

$$N_2O + N_2O^- \text{ (ads)} \rightarrow N_2 + O_2 + e^- \text{ (to catalyst)} \qquad (4)$$

For the examination of the possibility of these reactions, the reaction rates were examined kinetically by using the values of θ_V, $q^{re}(N_2O)$ and $q^{re}(O_2)$ which have been estimated from the results of transient response measurements. Since there are three possible reaction routes (1), (3) and (4), the reaction rates as measured by the formation rates of nitrogen can be tested by checking the fitness of rate equations developed for these different reaction routes. These rate equations are :

$$r_{decomp.} \text{ (1)} = k_1 P_{N_2O} \cdot \theta_V \qquad (5)$$

$$r_{decomp.} \text{ (3)} = k_3 P_{N_2O} \cdot q^{re}(O_2) \qquad (6)$$

$$r_{decomp.} \text{ (4)} = k_4 P_{N_2O} \cdot q^{re}(N_2O) \qquad (7)$$

In Figure 5 were shown the relationships between $r_{decomp.}$ and P_{N_2O} and $P_{N_2O} \cdot \theta_V$. A fairly good linear relation between $r_{decomp.}$ and $P_{N_2O} \cdot \theta_V$ is seen and this will also support the above conclusion that the decomposition of nitrous oxide proceeds though reaction (1). If reaction (3) or (4) is taking place, judging from the relationship between $r_{decomp.}$ and P_{N_2O} in Figure 5, $q(O_2)$ and $q(N_2O)$ should decrease with increasing P_{N_2O}. But this is not the case as can see from Figure 4. This will exclude the possibility of reaction (3) and (4).

It will, therefore, be plausible to conclude that the decomposition of nitrous oxide on CuO is taking place predominantly through the reaction step (1) with the desorption of oxygen as the slowest step.

MgO Catalyst. A charge of 42.6 g MgO catalyst was used in this experiment and kinetic measurements were made at 409°C. The response of a MgO catalyst equilibrated with helium stream to a step inflow of nitrous oxide was followed by monitoring N_2 and O_2 in the downflow stream and the results were presented in Figure 6.

As can be seen from the figure, the shape of the response curve of nitrogen is quite different from that obtained with a CuO catalyst, which showed a typical overshoot-type response. In the case of MgO, on the contrary, N_2 response showed a typical monotonic increasing response in the same fashion with that of oxygen. In this case, however, the time required to attain steady state is much shorter than in the case of CuO. Taking into account the fact that nitrogen can not be adsorbed on the catalyst surface, the lag in attaining steady state would suggest the slow adsorption of nitrous oxide. The ratio of evolved nitrogen and oxygen was approximately equal to the stoichiometric ratio indicating that the oxygen can scarcely be adsorbed on the surface or the adsorbed amount must be very small if any.

Figure 7 represents the results of $N_2O(dec.,0)-N_2,O_2,N_2O$

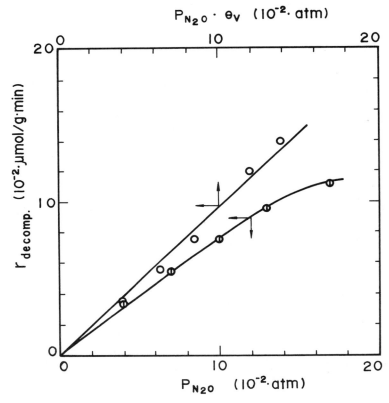

*Figure 5.   Relationship between the rate of decomposition and $P_{N_2O} \cdot \theta_V$ or P_{N_2O}
at 256°C.*

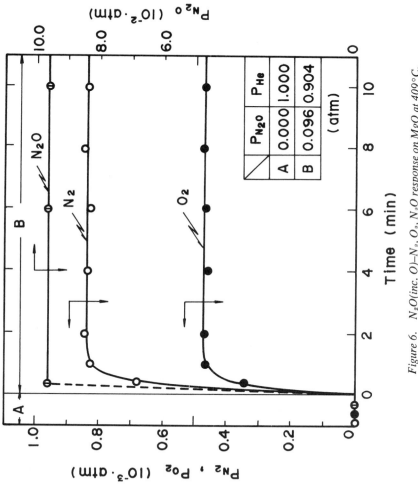

Figure 6. $N_2O(inc. O)-N_2, O_2, N_2O$ response on MgO at 409°C.

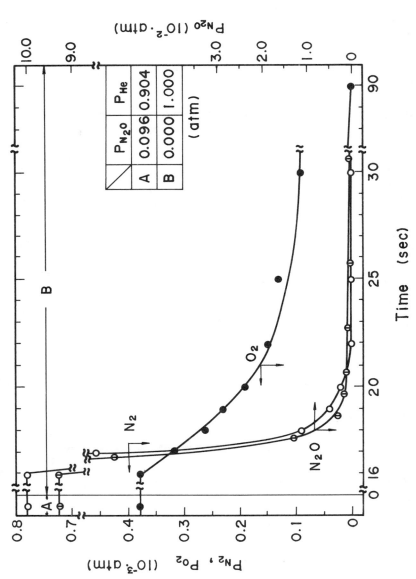

Figure 7. $N_2O(dec. O)$—N_2, O_2, N_2O response on MgO at 409°C.

response. Transient periods for these species were extremely short compared to those on CuO and this is clearly indicating the very small adsorption amounts of these species on the MgO catalyst in the stationary state of the reaction. It should be worth noted in this case that the evolution of nitrogen was observed even without nitrous oxide in the gas phase and that the evolution of both nitrogen and nitrous oxide ceased simultaneously. This fact strongly suggests that nitrous oxide decomposes after once adsorbed on the surface.

Since nitrous oxide was cut off from the feed stream, the sum of evolved nitrous oxide and nitrogen is equal to the adsorbed amount of nitrous oxide on the catalyst in stationary state of the reaction. This amount is extremely small compared to that on CuO and this fact also implys that the adsorption of nitrous oxide could be the slowest step in the overall reaction of nitrous oxide decomposition on MgO.

A linear relationship was observed between P_{N_2O} and the rates of decomposition as is shown in Figure 8.

The values of $q^{re}(N_2O)$ were estimated from the $N_2O(dec.,0)-N_2O$ response curve and were plotted against $r_{decomp.}$ as shown in Figure 9. As can be seen from the figure, the data were plotted almost in a linear form and the slope is approximately equal to 2.

These kinetic relationships suggest the following reaction mechanism including the adsorption of nitrous oxide as the slowest step.

$$N_2O + e^- \text{ (from catalyst)} \rightleftarrows N_2O^- \text{ (ads)} \qquad (8)$$

$$2 N_2O^- \text{ (ads)} \rightarrow 2 N_2 + 2 O^- \text{ (ads)} \qquad (9)$$

$$2 O^- \text{(ads)} \rightleftarrows O_2 + 2 e^- \text{ (to catalyst)} \qquad (10)$$

Discussion

Reaction mechanisms which are capable of explaining the experimental results on CuO and MgO are represented by Equations (1) and (2) for CuO and Equations (8),(9) and (10) for MgO.

In the experiments of transient response on CuO catalyst, it was found that no nitrogen was produced even with the adsorbed nitrous oxide still present on the surface. Although direct evidence was not provided in this study, based on the results and discussion presented by London and Bell (9), we inclined to believe that nitrous oxide adsorbed on sites through its nitrogen end will meet the steric problems in splitting N-O bond to produce nitrogen leaving an oxygen atom on the surface. When nitrous oxide hit the surface sites with its oxygen, on the other hand, this configuration does not meet such steric difficulties but may lead to the rapid breaking of N-O bond by facilitating the addition of electron from the catalyst which is necessary for weakening the bond due to the higher electronegativity of oxygen.

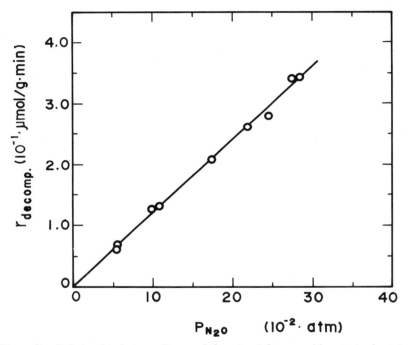

Figure 8.   Relationship between P_{N_2O} and the rate of decomposition at steady state on MgO at 412°C.

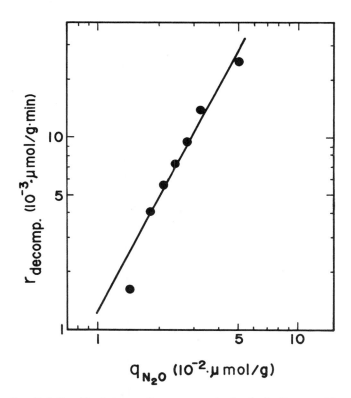

Figure 9.   Relationship between the amount of adsorbed nitrous oxide and the rate of decomposition at steady state on MgO at 409°C.

The faster desorption of nitrous oxide than oxygen as shown in Figure 1 may also be explained by this interpretation because the adsorption of nitrous oxide through its nitrogen end would be weaker than adsorption through its oxygen end as had been pointed out by Zecchina, Cerruti and Borello for adsorption of N_2O on Chromia (10).

London and Bell (9) have concluded in their study on the reaction between NO and CO over copper oxide that Cu^+ rather than Cu acts as the receipient of the oxygen atom produced by the dissociation of nitrogen oxide while Cu serves as the adsorption sites for nitrogen oxide through nitorgen end. In our experiments, however, there was no reducing reagent such as CO in their experiments, and hence the oxidation state of the catalyst would be different from that in their experiments. Therefore the further investigation would be necessary to see whether or not the valency state of copper plays a role in acting either as an active site for the decomposition of nitrous oxide or just as a site for the adsorpiton of nitrous oxide.

Relatively little has been done about the kinetics and mechanism of this reaction on MgO. Cimino and Indovina (11) and Cimino and Pepe (12) have investigated the decomposition of nitrous oxide on transient metal ions dispersed in MgO and suggested the possible role of the MgO matrix surface in releasing oxygen. They drew attention to the peroxide ion, which had been known to be formed on MgO and was posturated by Dowden (13) and by Birkley and Stone (14) to be responsible for the oxygen migration on the surface which leads to the desorption of oxygen. However, little has been discussed about the contribution of MgO surface to the adsorption or the decomposition step of nitrous oxide.

Vijh (6) has suggested more recently that if one assumes the adsorbed species formed in this reaction to be a covalent one, the available data can be interpreted in terms of the Sabatier-Balandin views on heterogeneous catalysis. According to his interpretation, he has indicated a volcanic relationship between the catalytic activity (defind as the temperature at which the reaction first becomes appreciable) against the heat of formation per equivalent of the oxide catalyst, ΔH_e. Based on this volcanic relationship, he has concluded that the rate-determining step (r.d.s.) of the reaction on the oxide catalysts such as CiO, NiO and CoO probably involves rupture of a M-O bond. On the other hand, r.d.s. on oxides such as MgO, CaO and CeO_2 would involve the formation of a M-O bond.

The desorption of oxygen which was suggested as the slowest step on CuO catalyst is the rupture of a M-O bond in its nature and the adsorption of nitrous oxide through oxygen end which was suggested as the slowest step on MgO catalyst is the formation of a M-O bond. Therefore, the results we have obtained in the present study appears to be consistent to the conclusion proposed by Vijh.

As has been discussed by Vijh (15) and Vijh and Lenfant (16), the activity seems to be related volvanically to the band gaps of the oxides and this would tend to emphasize the importance of semiconductivity as has been suggested by previous investigation (4).

Conclusions

The decomposition of nitrous oxide on CuO proceeds by the following reaction steps involving the desorption of oxygen as the slowest step.

$$N_2O + e^- \text{ (from catalyst)} \rightarrow N_2 + O^- \text{ (ads)}$$

$$2 \ O^- \text{ (ads)} \rightleftarrows O_2 + 2 \ e^- \text{ (to catalyst)}$$

The reaction on MgO, on the other hand, proceeds by the following mechanism involving the adsorption of nitrous oxide as the slowest step.

$$N_2O + e^- \text{ (from catalyst)} \rightarrow N_2O^- \text{ (ads)}$$

$$2 \ N_2O^- \text{ (ads)} \rightarrow 2 \ N_2 + 2 \ O^- \text{ (ads)}$$

$$2 \ O^- \text{(ads)} \rightleftarrows O_2 + 2 \ e^- \text{ (to catalyst)}$$

The mechanisms suggested in this study are consistent with that proposed by Vijh.

Literature Cited

1. Schwab, G. M. Z. Phys. Chem. 1938 25, 411, 418.
2. Wagner, C.; Hauffe, K. J. Chem. Phys. 1950, 18, 69.
3. Schmid, G. ; Keller, N. Naturwissenschaften 1950, 37, 43.
4. Dell, R. M. ; Stone, F.S.; Tiley, P. F. Trans. Faraday Soc. 1953, 49, 201.
5. Winter, E. R. S. Advance in Catal. 1958, 10, 232.
6. Vijh, A. K. J. Catal. 1973, 31, 51.
7. Kobayashi, H.; Kobayashi, M. Cat. Rev.-Sci. Eng. 1974, 10, 39.
8. Kobayashi, M.; Kobayashi, H. J. Catal. 1972, 27, 100.
6. London, J. W.; Bell, A. T. J. Catal. 1973, 31, 96.
10. Zecchina, A.; Cerrti, I.; Borello, F. J. Catal. 1972, 25, 55.
11. Cimino, A.; Indovina, V. J. Catal. 1970, 17, 54.
12. Cimino, A.; Pepe, F. J. Catal. 1972, 25, 362.
13. Dowden, D. A. Catal. Rev.-Sci. Eng. 1971, 5, 1.
14. Birkley, R. I.; Stone, F. S. Trans. Faraday Soc. 1968, 64, 3393.
15. Vijh, A. K. J. Catal. 1973, 28, 329.
16. Vijh, A. K.; Lenfant, P. Can. J. Chem. 1971, 49, 809.

RECEIVED July 28, 1981.

Transient and Steady-State Vapor Phase Electrocatalytic Ethylene Epoxidation

Voltage, Electrode Surface Area, and Temperature Effects

MICHAEL STOUKIDES and COSTAS G. VAYENAS

Massachusetts Institute of Technology, Department of Chemical Engineering, Cambridge, MA 02139

The catalytic activity and selectivity of polycrystalline silver catalysts used for ethylene epoxidation can be affected significantly by electrochemical O^{2-} pumping. This new phenomenon was studied in the solid electrolyte cell

$$C_2H_4, C_2H_4O, CO_2, O_2, Ag \mid ZrO_2(Y_2O_3) \mid Ag, air$$

at temperatures between 300° and 420° and atmospheric pressure. In the present communication we report on the effects of voltage, temperature and electrode surface area on the transient and steady-state behavior of the system. The change in the rate of C_2H_4O production can exceed the rate of O^{2-} pumping by a factor of 400 and is proportional to the anodic overvoltage both at steady state and during transients.

The catalytic epoxidation of ethylene on silver has been studied extensively over the last thirty years. The literature in this area is very broad and has been reviewed by several authors (1,2,3). In recent years considerable progress has been made towards a satisfactory understanding of the mechanism of this important and complex catalytic system.

Force and Bell have studied the infrared spectra of species adsorbed on silver during ethylene oxidation (4) and examined the relationship of these species to the reaction mechanism (5). Several authors have studied the effects of catalyst support and crystal size on activity and selectivity (6,7). Cant and Hall (8) used ^{14}C to study oxygen exchange between ethylene and ethylene oxide and found that C_2H_4O oxidation is much slower than direct CO_2 formation from ethylene oxidation in agreement with independent kinetic studies (1,6,12). Recent UHV studies have shown conclusively that CO_2 is the only product of C_2H_4 reaction with atomic oxygen (9) in agreement with previous studies on the interaction of N_2O and ethylene on silver (10). Kinetic measurements in a well mixed reactor have been recently combined with simultaneous in situ measurement of the thermodynamic activity of oxygen on

0097-6156/82/0178-0181$07.00/0
© 1982 American Chemical Society

porous silver catalyst films ($\underline{11},\underline{12}$). These results showed that thermodynamic equilibrium is not generally established between gaseous and surface oxygen during reaction and provided some additional evidence for the existence of at least two active adsorbed oxygen species ($\underline{12}$) in agreement with previous work ($\underline{2}$). However even one type of adsorbed oxygen only suffices to explain many of the existing results ($\underline{5}$) and there is considerable uncertainty about the exact nature of surface and subsurface oxygen on silver during reaction ($\underline{2},\underline{9}$). There is some evidence for surface oxide formation from the work of Seo and Sato ($\underline{13}$) who observed a continuous exo-electron emission during reaction which parallels the rate of ethylene oxide production.

Despite the existing uncertainty about the exact nature of the oxygen species present during reaction several different approaches have been reported which lead to an increase in the selectivity to ethylene oxide. These include alloying silver with other metals ($\underline{14}$) and using different catalyst supports and various promoters ($\underline{6},\underline{15}$). It is also known and industrially proven that the addition of few PPM of chlorinated hydrocarbon "moderators" to the gas feed improves the selectivity to ethylene oxide but decreases the catalyst activity ($\underline{15}$). It has also been found recently by Carberry et al that selectivity increases with γ-irradiation of the catalyst ($\underline{16}$).

In a recent communication a new electrochemical technique was described for altering the intrinsic activity and selectivity of silver for ethylene epoxidation ($\underline{17}$). The new technique utilizes a stabilized zirconia solid electrolyte cell with a porous silver film electrode exposed to ethylene/O_2 mixtures and functioning both as an electrode and as a catalyst for ethylene epoxidation (Fig. 1). When the circuit is open the silver film acts as a regular catalyst for ethylene epoxidation. When the circuit is closed and an external voltage V is applied to the oxygen concentration cell, O^{2-} is pumped electrochemically to the silver catalyst through the zirconia electrolyte. The activity of oxygen species on silver can thus be altered dramatically. This was shown to cause a significant change in catalyst activity and selectivity ($\underline{17}$). It was found that both catalyst activity and selectivity increase when O^{2-} is pumped to the catalyst and decrease when O^{2-} is electrochemically pumped from the catalyst by reversing the applied voltage. It was also found that the increase in the rate of ethylene oxide production can exceed the rate of O^{2-} transport through the electrolyte by a factor of 400. This increase varies linearly with applied current i for low values of current density ($<30\ \mu A/cm^2$) and becomes independent of i for current densities exceeding roughly $50\ \mu A/cm$. The relaxation time constants of the system showed conclusively that the new phenomenon is due to some type of surface rather than bulk oxidation of the silver crystallites ($\underline{17}$).

In the present communication we establish the effect of cell voltage, temperature and catalyst-electrode surface area on the

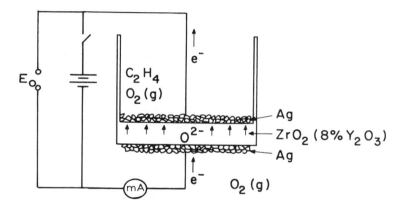

Figure 1.   Schematic diagram of the cell reactor.

transient and steady-state behavior of the system. It will be shown that the increase in the rates of ethylene oxide and CO_2 production is proportional to the cell overvoltage ΔV both during transients and at steady state. The implications of this observation upon the reaction mechanism will be discussed.

Zirconia cells similar to the ones employed in the present study, have been used i) by Mason et al (18) to electrochemically remove oxygen from Pt and Au catalysts used for NO decomposition. It was shown that electrochemical oxygen pumping causes a dramatic increase in the rate of NO decomposition (18,19), ii) by Farr and Vayenas to electrochemically oxidize ammonia and cogenerate NO and electrical energy (20,21), iii) by Vayenas et al (11,12,22,23) to study the mechanism of several metal catalyzed oxidations under open circuit (potentiometric) conditions.

The anodic oxidation of ethylene in a low temperature aqueous electrolyte electrochemical cell has been studied by Holbrook and Wise (24). It was found that product selectivity (carbonate, vs. ethylene glycol) depends on the potential of the silver anode.

Experimental Methods

The experimental apparatus has been described in detail elsewhere (11,12,22). In previous communications we have also described the porous silver catalyst film deposition and characterization procedure (11,12). Ten different reactor-cells were used in the present investigation. The cells differed in the silver catalyst surface area as shown in Table I. Catalysts 2 through 5 had been also used in a previous study (17). The reactor-cells also differed in the zirconia electrolyte thickness which could not be measured accurately. The electrolyte thickness varies roughly between 150 and 300 μm.

All reactor-cells used had a volume of 30 ml and have been shown to be well mixed over the range of flowrates employed in the present study (22). Both external and internal mass and heat transfer limitations have been shown to be negligible (12,22). Reactants were certified standards of ethylene diluted in N_2 and Matheson zero grade air. They could be further diluted in N_2. Reactants and products were analyzed by one line Gas Chromatography. The carbon dioxide concentration in the product stream was also continuously monitored using a non-dispersive IR CO_2 Analyzer (Beckman 864).

All runs reported here were done galvanostatically, i.e. by maintaining the imposed current constant and monitoring the cell voltage. Constant currents were applied to the cell using an AMEL 594 galvanostat.

Table I

Catalyst-electrode surface areas

Reactor - Cell #	Reactive oxygen uptake Q (moles O_2)
RC 2	$2.3 \cdot 10^{-7}$
RC 3	$.4 \cdot 10^{-7}$
RC 4	$2.5 \cdot 10^{-7}$
RC 5	$20. \cdot 10^{-7}$
RC 6	$6.5 \cdot 10^{-7}$
RC 7	$9.4 \cdot 10^{-7}$
RC 8	$5.2 \cdot 10^{-7}$
RC 9	$.4 \cdot 10^{-7}$
RC 10	$10. \cdot 10^{-7}$
RC 11	$2.6 \cdot 10^{-7}$

Results

The pure anionic conductivity of the zirconia electrolyte was verified by passing through the reactor-cell mixtures of O_2, N_2 and He of known P_{O_2} and observing close agreement (± 2mV) between measured and theoretical emf values. However when ethylene /O_2 mixtures are fed to the reactor the open circuit emf E which reflects the oxygen activity a_O^2 on the catalyst (11,12) according to

$$E = \frac{RT}{4F} \ln a_O^2 /(.21) \qquad (1)$$

takes values between -10 and -80 mV, indicating that $a_O^2 < P_{O_2}$. This implies that thermodynamic equilibrium is not established in general between gaseous and adsorbed oxygen during reaction, because the intrinsic rate of oxygen adsorption is comparable to the rate of the surface oxidation step (11,12).

When a freshly calcined silver catalyst is placed on stream in an ethylene/O_2 mixture the catalyst activity and selectivity as well as the surface oxygen activity a_O change considerably during the first 10 h and approach steady state only after 30 h on stream. The selectivity usually passes through a maximum. A typical case is shown in figure 2. The catalyst activity, selectivity and oxygen activity were then found to remain practically constant over periods of several weeks. All the oxygen pumping experiments

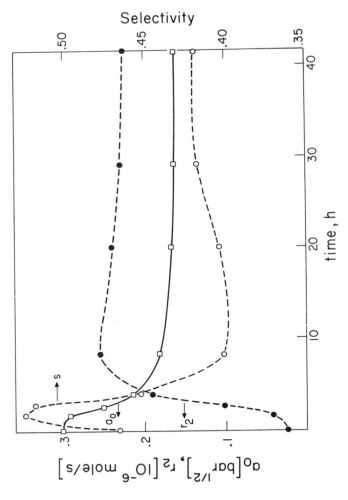

Figure 2. Transient catalytic activity, selectivity, and surface oxygen activity of a freshly calcined silver catalyst during the first 40 h on stream. Catalyst is exposed to reactive gas mixture at $t = 0$, $410°C$, $P_{ET} = 0.015$ bar, and $P_{O_2} = 0.10$ bar.

reported here were performed after the silver catalysts had
reached steady state.

The effect of electrochemical oxygen pumping on the rates of
ethylene epoxidation r_1 and deep oxidation r_2 is shown in figure
3. At time t<o the circuit is open and the catalyst is at a
steady state characterized by r_{10} = 7.6·10^{-8} mole/s and r_{20} = 8.9·
10^{-8} mole/s. At t = o a constant current of +25 mA is applied to
the cell and O^{2-} is pumped electrochemically to the catalyst at a
rate $G_{O^{2-}}$ = i/2F = 1.3·10^{-10} mole/s. If the applied voltage had
no effect on the Ag catalyst and each O^{2-} pumped to the catalyst
yielded one C_2H_4O molecule then the maximum expected relative in-
crease $\Delta r_1/r_{10}$ is $G_{O^{2-}}/r_{10}$ = 1.7·10^{-3}. As figure 3 shows $\Delta r_1/r_{10}$
= .2, i.e. Δr_1 is two orders of magnitude larger than $G_{O^{2-}}$, indicating a dramatic change in the properties of the Ag catalyst.
During the transient as well as at steady state the ratio $\Delta r_1/\Delta r_2$
remains practically constant at 1.7. At time t = 90 min the circuit is opened and the catalyst returns to its initial state indicating the reversibility of the phenomenon. At t = 180 min a constant current of -25 μA is applied through the cell so that oxygen
is pumped from the catalyst. The rates r_1, r_2 as well as the selectivity drop. When the circuit is opened again the rates return
to the original steady state values r_{10} and r_{20}.

Overvoltage effects: Figure 4 shows a typical transient cell
voltage response when a constant current i is imposed to the cell
at time t = 0. At t<o the cell voltage equals the open circuit
emf E given by (1). The initial rapid rise at t = 0 is caused by
the ohmic drop iR_c, where R_c is the ohmic resistance of the cell
which is primarily due to the zirconia electrolyte resistance. The
subsequent gradual increase ΔV to a final asymptotic value, corresponds to charging of the electrode-electrolyte double layers and
formation of surface intermediates on the silver electrode (25).
These surface intermediates must be adsorbed oxygen species, possibly surface silver oxide (17). The increase ΔV is the cell
overpotential; ΔV as well as the relaxation time constant τ_o,
defined as the time required for ΔV to reach 63% of its final
steady state value (17,25) were both found to be strongly dependent on gas phase composition. They both vanish as P_{O_2}/P_{ET} approaches zero. It can thus be concluded that double layer charging of the electrode-electrolyte interface happens quite fastly
and has a negligible contribution to the ΔV transient.

Within the accuracy of the experimental data the galvanostatic transient response of ΔV is identical to the transient rate response Δr_1 and Δr_2, i.e. $\tau_o = \tau_c$ where τ_c is the relaxation time
constant for the two rates (17). This is shown in figure 5 for
two different reactors under similar operating conditions and also
in figure 6 where the transient and the steady state Δr_i values
from four reactors are plotted vs. the cell overvoltage ΔV. In
view of the fact that r_{io} is proportional to the surface area Q it
follows from figure 6 that for constant gas phase composition

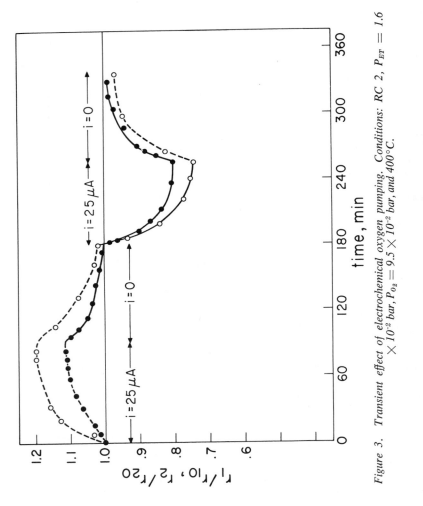

Figure 3. Transient effect of electrochemical oxygen pumping. Conditions: RC 2, $P_{ET} = 1.6 \times 10^{-2}\ bar$, $P_{O_2} = 9.5 \times 10^{-2}\ bar$, and 400°C.

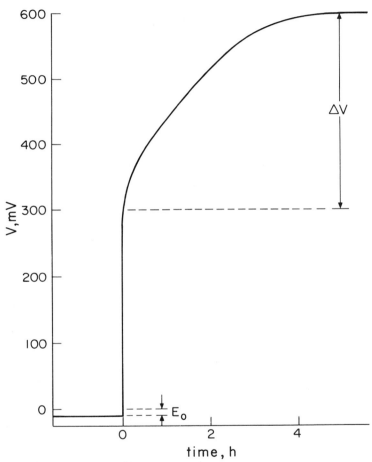

Figure 4.   Transient cell voltage response when a constant current i = +100 μA is applied at t = 0, P_{O_2} = 0.140 bar, and P_{ET} = 0.008 bar. Conditions: RC 8, i = 100 μA, and 400°C.

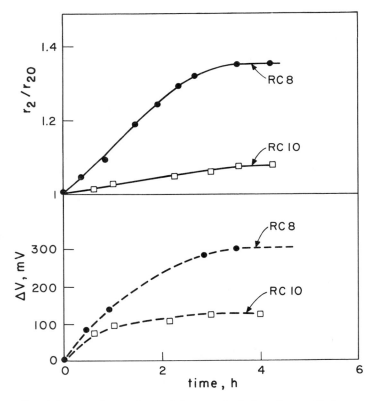

Figure 5. Galvanostatic transients of the rate of ethylene deep oxidation r_2 and of the cell overvoltage. The rate of epoxidation r_1 parallels r_2. Conditions: $i = 100 \mu A$, $P_{O_2} = 0.14$ bar at $400°C$. Key: ●, RC 8, $P_{ET} \approx 0.008$ bar; □, RC 10, $P_{ET} \approx 0.0068$ bar.

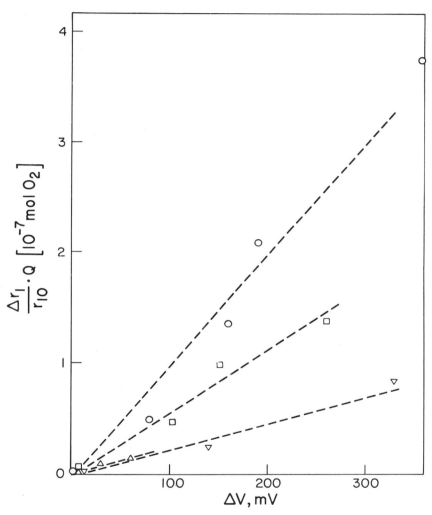

Figure 6.   Overvoltage effect on the increase in the rate of ethylene epoxidation at constant gas phase composition and $i = 100$ μA. Key: $\bigcirc$, RC 7, $P_{O_2}/P_{ET} \approx 21.0$, 400°C; $\square$, RC 8, $P_{O_2}/P_{ET} \approx 12.1$, 400°C; $\bigtriangledown$, RC 9, $P_{O_2}/P_{ET} \approx 7.9$, 420°C; and $\triangle$, RC 10, $P_{O_2}/P_{ET} \approx 6.8$, 400°C.

$$\Delta r_i \; \alpha \; \Delta V \tag{2}$$

Equation (2) is valid both during transients as well as at steady state in which case

$$(\Delta r_i)_{max} \quad \alpha \; \Delta V_{max}$$

As noted earlier the increase in the rates Δr_i can exceed the rate of oxygen transport through the electrolyte $i/4F$ by more than two orders of magnitude (Fig. 7).

Gas phase composition effects: As shown in figure 6 the increase in the rates Δr_i depends not only on the overvoltage ΔV but also on gas phase composition. This has been reported also in a previous communication where $\Delta r_i/r_{io}$ as well as τ_c were found to increase with decreasing P_{ET} (17). A series of runs performed at constant P_{ET} showed that $\Delta r_i/r_{io}$ is proportional to P_{O_2}. It was then established that $\Delta r_i/r_{io}$ is proportional to P_{O_2}/P_{ET} and that all the data, transient and steady state, obtained with ten different reactors, each having different electrode surface area can be correlated in terms of one surprisingly simple equation,

$$\frac{\Delta r_i}{r_{io}} \cdot Q \cdot \frac{P_{ET}}{P_{O_2}} = \alpha_i \Delta V \tag{3}$$

where α_1, α_2 (i=1,2) are constants with a ratio $\alpha_1/\alpha_2 = 1.65 \pm .3$. This is shown in figure 8. Equation (3) contains as a limiting case equation (2) which is valid for constant gas phase composition.

It should be noted that at temperatures above 350°C r_{io} is first order in ethylene and zero order in oxygen (12), i.e.

$$r_{io} = k_{io} \cdot Q \cdot P_{ET} \tag{4}$$

It thus follows that equation (3) can be written as

$$\Delta r_i = k_{io} \cdot \alpha_i \cdot P_{O_2} \cdot \Delta V \tag{5}$$

which contains as a limiting case the observation that Δr_i vanishes when $P_{O_2} \to 0$, i.e. when the reactor is run as a fuel cell (17). Equations (3),(4) and (5) are also valid both during transients and at steady state.

Most galvanostatic transients followed a first order response with reasonable accuracy, in agreement with (25):

$$\Delta V = \Delta V_{max}(1 - e^{-t/\tau_o}) \tag{6}$$

Table II shows the dependence of τ_o on gas phase composition. For constant surface area and current the time constant τ_o is roughly proportional to $\Delta V_{max} \cdot P_{O_2}/P_{ET}$. It is also shown that τ_o

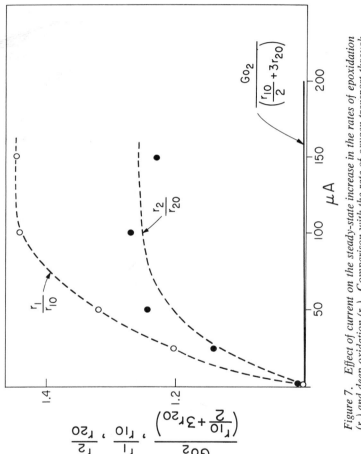

Figure 7. Effect of current on the steady-state increase in the rates of epoxidation (r_1) and deep oxidation (r_2). Comparison with the rate of oxygen transport through the electrolyte $G_{O_2} = i/4F$. Intrinsic selectivity $r_{10}/r_{20} = 0.49$. Conditions: RC 2 at 400°C, $P_{ET} \simeq 0.016$ bar, $P_{O_2} \simeq 0.1$ bar.

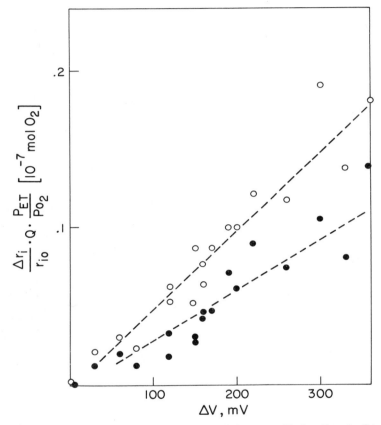

Figure 8.   Overvoltage effect on the rates of ethylene epoxidation (i = 1, ○) and deep oxidation (i = 2, ●).   At constant overvoltage the relative rate increases are proportional to P_{O_2}/P_{ET} at 400°C.

is proportional to $\Delta r_i/r_{io}$ in agreement with equation (3). The decrease in $\tau_c \sim \tau_o$ with increasing P_{ET} has been also reported previously (17).

Table II

Gas phase composition effect on
the relaxation time constant τ_o for i=100 A and T=400°C

Reactor #	Q	P_{O_2}/P_{ET}	ΔV_{max}	r_1/r_{10}	r_2/r_{20}	τ_o
	moles O_2		mV			min
RC 8	$5.2\cdot10^{-7}$	17.5	300	.64	.35	150
RC 8	$5.2\cdot10^{-7}$	12.1	250	.27	.17	75
RC 8	$5.2\cdot10^{-7}$	7.9	180	.15	.09	35
RC 10	$10\cdot10^{-7}$	6.9	60	.02	.015	20
RC 10	$10\cdot10^{-7}$	20.5	160	.16	.085	130

Catalyst-Electrode Surface Area Effects: At constant temperature, gas phase composition and current, therefore constant overvoltage, the relative increase in the reaction rates $\Delta r_i/r_{io}$ is inversely proportional to the electrode surface area. This is a consequence of equation (3) and is shown in figure 9 for five different reactors. Thus in order to maximize the relative effect of oxygen pumping on r_1 and r_2, i.e. in order to maximize yield and selectivity of ethylene oxide one must minimize the electrode surface area, i.e. minimize the porous silver film thickness without causing a major decrease in its conductivity. A decrease in the electrode surface area also causes a moderate decrease in the relaxation time constant τ_c. This is shown in figure 10 which contains data from 5 reactors obtained at $P_{O_2}/P_{ET} \sim 7$ and T = 400°C and shows that $1/\tau_o$ is a linear increasing function of i/4FQ. An explanation for these observations is given in the discussion section.

Figure 11 demonstrates a problem frequently encountered during the oxygen pumping experiments. The galvanostatic transient was obtained with reactor RC9 which had the lowest electrode surface area Q(Table 1), therefore exhibited the largest relative effect $\Delta r_i/r_{io}$, i.e. a threefold increase in ethylene oxide yield and a 210% increase in CO_2 production with a corresponding 20% increasing in selectivity. As shown on the figure the constant current i = 100 μA was applied at t=o. At t=40 min and long before the reactor would reach steady state the cell resistance suddenly dropped from a few KΩ to 1-2 Ohms indicating electrolysis of the zirconia electrolyte and metallic zirconium formation. This is possible since the total cell voltage ($\Delta V+iR_c$) at t=40 min had reached approximately 2.5 V. As shown on the figure subse-

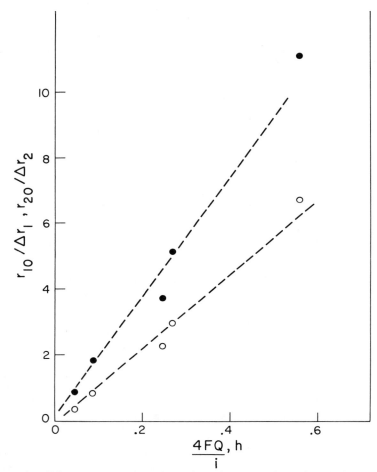

Figure 9.   Effect of Ag catalyst–electrode surface area Q on the relative steady-state increase in the rates of epoxidation r_1 and deep oxidation r_2 at constant imposed current $i = 100$ μA, constant gas composition, $400°C$, $P_{O_2}/P_{ET} \sim 7$. Key: $\bigcirc$, $r_{10}/\Delta r_1$; and $\bullet$, $r_{20}/\Delta r_2$.

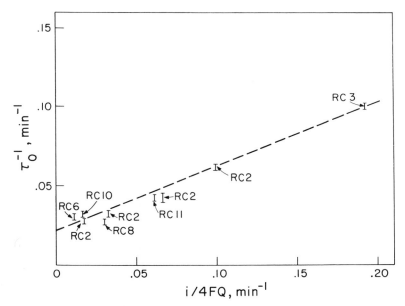

Figure 10.   Effect of Ag catalyst–electrode surface area Q on the cell relaxation time constant. Conditions: $400\,^{\circ}C$, $P_{O_2}/P_{ET} \simeq 7$.

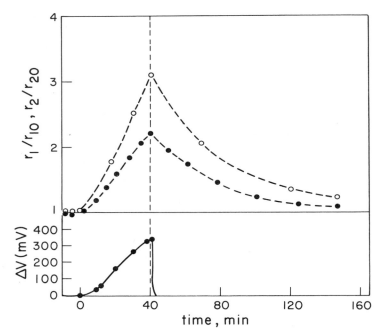

Figure 11.   Transient galvanostatic response of the rates of ethylene oxidation r_1 (○) and deep oxidation r_2 (●) and of the cell overvoltage ΔV for reactor RC 9. Electrolyte breakdown occurred at 40 min.   Conditions: RC 9, 420°C, i = 100 μA, $P_{O_2} = 0.095$ bar, $P_{ET} = 0.012$ bar.

quently to the electrolyte breakdown both r_1 and r_2 slowly return towards their intrinsic values r_{10} and r_{20} since electronic rather than ionic conduction is now occuring across the cell. The cell cannot function as an oxygen concentration cell anymore since the electrolyte has been destroyed; the open circuit emf E always remains equal to zero regardless of the reactor gas composition. It was thus concluded that total voltages in excess of 2.5 V should not be applied to the cell at these relatively low temperatures.

Temperature effects: At constant current and gas composition, temperature has an interesting effect on the steady state rate increases Δr_1 and Δr_2 (Fig. 12). At temperatures below 350°C both Δr_1 and Δr_2 increase rapidly with an activation energy E*=33+3 Kcal/mole. However at temperatures above 350°C Δr_1 and Δr_2 decrease with an apparent negative activation energy E** = 19 + 2 Kcal/mole. The overvoltage ΔV decreased monotonically from 90 mV at 320°C to 3–5 mV at 400°C. It thus appears that the behavior of ΔV parallels the behavior of Δr_1 and Δr_2 at high T but not at temperatures below 350°C. A possible explanation is given in the discussion section. Figure 12 also shows that within the accuracy of the experimental data the ratio $\Delta r_1/\Delta r_2$ remains practically constant at 1.75+.25 which corresponds to .64+.04 selectivity on the oxide.

Discussion

The nature of the silver catalyst is clearly altered significantly during electrochemical oxygen pumping. The increase or decrease in the rates of ethylene epoxidation and combustion is more than two orders of magnitude higher than the rate of oxygen transport through the electrolyte. The selectivity also changes considerably.

When oxygen is pumped to the catalyst the activity of oxygen on the silver catalyst-electrode increases considerably because of the applied voltage. It thus becomes possible to at least partly oxidize the silver catalyst electrode. In a previous communication it has been shown that the phenomenon involves surface rather than bulk oxidation of the silver crystallites (17). The present results establish the direct dependence of the change in the rates of epoxidation and combustion Δr_1 and Δr_2 on the cell overvoltage (Equations 2,3, and 5) which is directly related to the surface oxygen activity.

Furthermore, the observation that the transient rate change behavior parallels the transient overvoltage behavior (Figures 5, 6 and equations 2,3 and 5) proves undoubtedly that the increase or decrease in the rates is caused by a corresponding increase or decrease in the surface concentration of an adsorbed oxygen species or surface silver oxide. In a previous communication we have denoted this unknown surface oxygen species by AgO_2^* with a concentration $c(moles/cm^2)$. There is previous evidence in the liter-

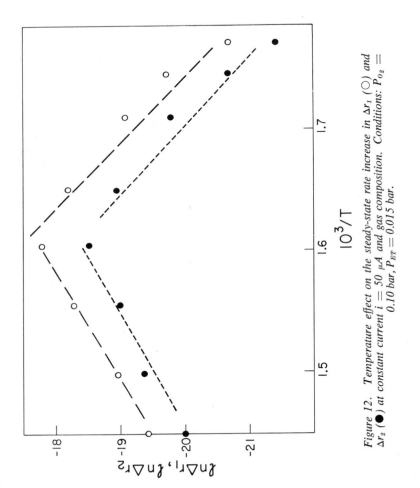

Figure 12. Temperature effect on the steady-state rate increase in Δr_1 ($\bigcirc$) and Δr_2 ($\bullet$) at constant current $i = 50$ μA and gas composition. Conditions: $P_{O_2} = 0.10$ bar, $P_{ET} = 0.015$ bar.

ature for the existence of surface silver oxides including the
work of Seo and Sato (13) who observed a continuous exoelectron
emission from silver catalysts during ethylene epoxidation which
was proportional to the rate of ethylene oxide formation. They explained their results in terms of formation of surface silver
oxide Ag_2O with molecular oxygen ion O_2^- adsorbed on it (13). This
picture seems to be in excellent agreement with the present results, i.e. with equation (5)

$$\Delta r_i = k_{io} \cdot \alpha_i \cdot \Delta V \cdot P_{O_2} \tag{5}$$

if the change in the amount of surface silver oxide is proportional to ΔV and the coverage of molecular oxygen adsorbed on the
oxide is proportional to P_{O_2}, or at least if the rates of epoxidation and combustion on the silver oxide are first order in
oxygen.
      Equations (2), (3) and (5) which describe the dependence of
Δr_i on the overvoltage are extremely simple but deserve some special attention. They all indicate that Δr_i is proportional to ΔV
but independent of the surface area of silver electrode. This is
why the relative increase in the rates $\Delta r_i/r_{i_o}$ is highest for reactors with small surface area (Fig. 11) and becomes insignificant
for reactors with large surface area. It thus appears that although ΔV is an intensive variable it is nevertheless a measure of
the change $S \cdot \Delta c$ in the mole number of silver oxide and not of the
concentration change Δc. This is because all the reactors used
had the same electrolyte area $A = 2cm^2$ as explained in detail below: Neglecting the double layer capacitance of the electrode
electrolyte interface the constant current i imposed during the
transient can be split in two parts (25):

$$i = C_{ad} \frac{d(\Delta V)}{dt} + i_{CTR} \tag{7}$$

where i_{CTR} is the current corresponding to the charge transfer reactions, i.e.

$$O^{2-} + C_2H_4 \rightarrow Products + 2e^- \tag{8}$$

$$O^{2-} \rightarrow 1/2\ O_2(g) + 2e^- \tag{9}$$

and C_{ad} is the adsorption pseudo capacitance of the oxide which by
definition equals (25):

$$C_{ad} = \frac{dq}{dV} = \lambda \cdot F \cdot \frac{d(Sc)}{dV} \tag{10}$$

where q is the charge stored in the oxide and λ is a constant. If
one assumes that Sc varies linearly with V, i.e. that the oxide
pseudo capacitance is constant one obtains

$$C_{ad} = \lambda \cdot F \frac{\Delta(Sc)}{\Delta V} \tag{11}$$

The capaciatance C_{ad} depends on the gas–electrode–electrolyte interline "area" S´ but not on the total electrode surface area S. If the porosity of all the electrode catalysts used is the same, which is a reasonable assumption since they were all prepared by the same calcination procedure, it follows that the interline "area" S´ is proportional to the flat electrolyte surface area A, i.e. the constant λ equals $\lambda_o A$, where λ_o is another constant which does not depend on any macroscopic dimension.

$$C_{ad} = \lambda_o \cdot A \cdot F \cdot \frac{\Delta(Sc)}{\Delta V} \tag{12}$$

Since all reactors used in the present study had the same electrolyte surface area A, it then becomes clear that they all had the same adsorption pseudo capcitance C_{ad}, therefore ΔV is proportional to $\Delta(Sc)$ as previously stated.

Quantitative kinetics: The experimental relations (2) through (5) which express the dependence of Δr_i and $\Delta r_i/r_{io}$ on ΔV and gas phase composition can now be explained in a quantitative manner in terms of a simple model which makes the same assumptions with those made in (17) but also takes into account that the change in the moles of silver oxide (Sc) caused by electrochemical oxygen pumping is directly measurable and proportional to the overvoltage ΔV. Assuming that the changes Δr_i in the rates of ethylene epoxidation r_1 and combustion r_2 are proportional to the change in the moles of AgO_2^* and that both reactions on the AgO_2^* surface are first order in oxygen one obtains

$$\Delta r_i = k_i \cdot P_{O_2} \cdot \Delta(Sc) \tag{13}$$

Using equation (12) one obtains

$$\Delta r_i = k_i \cdot \lambda_o \cdot A \cdot F \cdot (\Delta V) \cdot P_{O_2} \tag{14}$$

which coincides with the experimental equation (5) with $k_{io} \alpha_i = k_i \cdot \lambda_o \cdot A \cdot F$. It should be noted that according to (14) Δr_i is proportional to the electrolyte surface area A but does not depend on the electrode surface area. The experimental relation

$$\Delta r_i \propto \Delta V \tag{2}$$

valid for constant gas phase composition is obviously a special case of (14). Finally to explain the dependence of $\Delta r_i/r_{io}$ on ΔV and gas phase composition one must assume that oxide forms on silver sites which are inactive for reaction before the oxide forms and take into account that at temperatures above 350°C and under open–circuit conditions both r_{10} and r_{20} are first order in ethylene and zero order in oxygen (12).

$$r_{io} = k_{io} \cdot Q \cdot P_{ET} \tag{4}$$

This last equation is not inconsistent with the hypothesis that the rates on AgO_2^* are first order in oxygen to the extent that the amount of AgO_2^* is small under open circuit conditions.

Combining with equation (14) one obtains the experimental relation

$$\frac{\Delta r_i}{r_{io}} \cdot Q \cdot \frac{P_{ET}}{P_{O_2}} = \alpha_i \Delta V \tag{3}$$

where

$$\alpha_i = \frac{k_i}{k_{io}} \cdot \lambda_o \cdot A \cdot F \tag{15}$$

Surface area effect on the relaxation time constant: In a previous communication (17) we have developed a simple dynamic model which allows one to predict the change in the mole number of silver oxide $S \cdot \Delta c$, therefore ΔV, in terms of the imposed current, P_{O_2} and P_{ET}. The dynamic equation of the model was

$$S \cdot \theta_{O_2} \cdot \frac{dc}{dt} = \frac{i}{4F} (1-\frac{c}{c_M}) + K_1 \cdot S\theta_{O_2} \cdot (c_M-c) - K_2 \cdot S \cdot P_{ET} \cdot c \tag{16}$$

with $\theta_{O_2} = K_{O_2}P_{O_2}/(1+K_{O_2}P_{O_2})$ \hfill (17)

i.e. the model assumed that oxide can form only on sites covered originally with molecular O_2. According to this equation the relaxation time constant of the system $\tau_c = \tau_o$ is given by

$$\frac{1}{\tau_c} = \frac{4FSC_M}{i} (1+ \frac{1}{K_{O_2}P_{O_2}}) + K + \frac{K_2P_{ET}}{K_{O_2}P_{O_2}} (1+K_{O_2}P_{O_2}) \tag{18}$$

Equation (18) is in excellent agreement with figure 10 which shows $1/\tau_c$ for 5 different reactors to be a linear increasing function of $4FQ/i$ with a positive intercept. One of the implications is that the amount of surface oxide at full coverage Sc_M is comparable to the reactive oxygen uptake Q of the catalyst.

Temperature effect: There are several ways to account for the observed maximum in Δr_i at constant gas composition and imposed current i (Fig. 12). Although the behavior shown in the figure is quite reproducible, further experimental work is required to examine the effect of gas phase composition and i on the temperature dependence of Δr_i. One possible explanation for the observed maximum can be obtained by closer examination of equation (7). The charge transfer current i   can be written as $i_{CTR,A}$ + $i_{CTR,B}$ where the former term corresponds to the charge transfer reaction (8) and the latter to the charge transfer reaction (9). Using the low field law (25) one obtains

$$i_{CTR,A} = i_{0,A} \cdot F(\Delta V)/RT \tag{19}$$

$$i_{CTR,B} = i_{0,B} \cdot F(\Delta V)/RT \tag{20}$$

The exchange current densities $i_{0,A}$ and $i_{0,B}$ are temperature dependent and can be expressed as

$$i_{0,A} = \vec{K}_{0,A} \exp(-\frac{\Delta G_A^{\ddagger}}{RT}) \tag{21}$$

$$i_{0,B} = \vec{K}_{0,B} \exp(-\frac{\Delta G_B^{\ddagger}}{RT}) \tag{22}$$

where $\vec{K}_{0A}$, $\vec{K}_{0,B}$ are constants and $\Delta G_A^{\ddagger}$, $\Delta G_B^{\ddagger}$ are the standard Gibbs energies of formation of the activated complex for the charge transfer reactions (8) and (9).

Thus equation (7) can be written as:

$$C_{ad}\frac{d(\Delta V)}{dt} = i - \frac{F(\Delta V)}{RT}[\vec{K}_{0,A}\exp(-\frac{\Delta G_A^{\ddagger}}{RT})+\vec{K}_{0,B}\exp(-\frac{\Delta G_B^{\ddagger}}{RT})] \tag{23}$$

At steady state equation (23) gives

$$(\Delta V) = \frac{iRT}{F \cdot \vec{K}_{0,A} \cdot \exp(-\frac{\Delta G_A^{\ddagger}}{RT}) \cdot [1+\frac{\vec{K}_{0,B}}{\vec{K}_{0,A}}\exp\ (-(\frac{\Delta G_B^{\ddagger}-\Delta G_A^{\ddagger}}{RT})]} \tag{24}$$

If the activation energies for the epoxidation and combustion reactions on silver oxide equal E*, then the rate coefficients k_i in equations (14) can be expressed as

$$k_i = k_i^o\exp(-E*/RT) \tag{25}$$

Combining equations (14),(24) and (25) one obtains

$$\Delta r_i = \frac{k_i^o \cdot \lambda_o \cdot A.P_{O_2} \cdot i \cdot RT \cdot \exp(\frac{-E*+\Delta G_A^{\ddagger}}{RT})}{1 + \frac{\vec{K}_{0,B}}{\vec{K}_{0,A}} \exp\ (-\frac{(\Delta G_B^{\ddagger}-\Delta G_A^{\ddagger})}{RT})} \tag{26}$$

Equation (26) predicts a maximum in Δr_i for a large number of combinations of E*, $\Delta G_A^{\ddagger}$,$\Delta G_B^{\ddagger}$. As an example the set E* = -43 Kcal/mol, $\Delta G_A^{\ddagger}$ = 10 Kcal/mol and $\Delta G_B^{\ddagger}$ = 62 Kcal/mole yields the correct experimental values for the phenomenological activiation energies of Δr_i, i.e. +33 Kcal/mol at low T and -19 Kcal/mole at high T. It should be noticed however that these phenomenological activation energies must be gas phase composition dependent since the quantities $\vec{K}_{0,A}$, $\vec{K}_{0,B}$, $\Delta G_A^{\ddagger}$,$\Delta G_B^{\ddagger}$ which appear in equation (26) will depend on the coverages of ethylene and oxygen at the gas-electrode-electrolyte three phase boundary.

Conclusions

The activity and ethylene oxide selectivity of porous poly-crystalline silver catalysts can be altered significantly by using

the catalyst also as an electrode of a stabilized zirconia cell
and electrochemically pumping oxygen to or from the catalyst.
Oxygen pumping to the catalyst increases both activity and ethylene oxide selectivity. The opposite phenomenon is observed upon
reversing the voltage polarity. The increase or decrease in the
rate of ethylene oxide and CO_2 production can be at least two orders of magnitude higher than the rate of oxygen transport through
the electrolyte.

The relative increase $\Delta r_i/r_{i_o}$ in the rates of epoxidation
(i=1) and combustion (i=2) is proportional to A/S, where A is the
electrolyte surface area and S is the surface area of the silver
catalyst electrode. Thus with a reactor having a low value of S
(reactive oxygen uptake Q $= .4 \cdot 10^{-7}$ mol O_2) a threefold increase in
ethylene oxide yield was observed with a corresponding 20% increase in selectivity.

The absolute increase Δr_i in the rates of epoxidation and
combustion is directly proportional to P_{O_2} and the cell overvoltage ΔV both at steady state and during galvanostatic transients.
The ratio $\Delta r_1/\Delta r_2$ equals 1.7$\pm$ .3 and appears to be rather insensitive to current, overvoltage, gas phase composition and tempera-
ture.

The relaxation time constants of the overvoltage (τ_o) and the
rates (τ_c) are practically equal and of order 4FQ/i. They increase with increasing P_{O_2}/P_{ET} ratios. The new observations are
in reasonably good agreement with a previously proposed model (17).

The phenomenon appears to be due to formation and destruction
of some type of surface silver oxide during oxygen pumping to and
from the catalyst respectively. The use of in situ surface science techniques should prove very useful for the elucidation of
the exact nature of this surface oxide.

The present study was limited to temperatures above 300°C because of the severe increase in the electrolyte resistance with
decreasing temperature. Thinner (<100 µm) electrolyte components
must be used to study the temperature range of industrial interest,
i.e. 220° to 290°C. We note that this is the first known application of electrochemical oxygen pumping through a solid electrolyte
to alter the selectivity of a heterogeneous catalyst reaction. The
application of external voltages to solid electrolyte cells with
appropriate catalytic electrodes may be a powerful technique for
influencing the selectivity of other partial catalytic oxidations
as well.

List of symbols

a_o = Oxygen activity, $bar^{1/2}$

A = Electrolyte surface area, cm^2

c = Surface concentration, mol/cm^2

c_m = Surface concentration at full coverage, mol/cm^2

C_{ad} = Adsorption pseudo capacitance, Cb/V

E = Open circuit emf, V

E^* = Activation energy for reaction on the surface oxide, $Kcal/mol$

F = Faraday constant, $96500\ Cb/mol$

i = Current, A

i_{CTR} = Current of charge transfer reactions, A

$i_{O,A}, o_{O,B}$ = Exchange currents for reactions (8) and (9), A

k_i = Rate constant for ethylene epoxidation (i=1) and deep oxidation (i-2) on the surface oxide, $s^{-1} \cdot bar^{-1}$

k_{io} = Rate constant for ethylene epoxidation (i=1) and deep oxidation (i=2) on reduced silver, $s^{-1} \cdot bar^{-1}$

$\vec{k}_{O,A}, \vec{k}_{O,B}$ = Preexponential factors of the exchange current of reactions (8) and (9), A

K_1 = Rate coefficient, s^{-1}

K_2 = Rate coefficient, $s^{-1} \cdot bar^{-1}$

K_{O_2} = Molecular O_2 adsorption coefficient, bar^{-1}

Q = Reactive oxygen uptake, mols O_2; proportional to catalyst-electrode surface area S.

r_i = Rate of ethylene epoxidation (i=1) and deep oxidation (i=2), -mols ethylene/s

r_{io} = intrinsic open circuit rates of ethylene epoxidation (i=1) and deep oxidation (i=2), -mols ethylene/s

R = Gas constant, $8.31\ J/mol \cdot K$

R_c = Ohmic cell resistance, Ω

S = Catalyst electrode surface area, cm^2

t = time, s

V = Cell voltage, Volts

Greek symbols

α_i = Proportionality constants, mol/V

Δr_i = Change in rate of ethylene epoxidation (i=1) and deep oxidation (i=2) due to oxygen pumping, -mols ethylene/S

ΔV = Overpotential, Volts

θ_{O_2} = Molecular oxygen coverage, dimensionless

λ = Constant, dimensionless

λ_o = Constant, cm^{-2}

τ_c = Rate relaxation time constant, min

τ_o = Voltage relaxation time constant, min

Acknowledgements

This research was supported under NSF Grant CPE-8009436 and under the MIT J.R. Mares career development professorship Grant.

Literature Cited

1. Voge, H.H.; Adams, C.R. Adv. Catal. 1967, 17, 154

2. Kilty, R.A.; Sachtler, W.M.H. Cat. Rev.-Sci. Eng. 1974, 10(1) 1

3. Verykios, X.E.; Stein, F.P.; Coughlin, R.W. Catal. Rev. Sci. Eng. 1980, 22(2), 197

4. Force, E.L.; Bell, A.T. J. Catal. 1975, 38, 440

5. Force, E.L.; Bell, A.T. J. Catal. 1975, 40, 356

6. Wu, J.C.; Harriot, P. J. Catal. 1975, 39, 395

7. Verykios, X.E.; Stein, F.P.; Coughlin, R.W. J. Catal. 1980, 66, 368

8. Cant, N.W.; Hall, W.K. J. Catal. 1978, 52, 81

9. Wachs, I.E.; Keleman, S.R. Proc. 7th Inter. Congr. Catal. 1980, paper A48.

10. Herzog, W. Ber. Bunsen-Gesel. 1970, 74 (3), 216

11. Stoukides, M.; Vayenas, C.G. J. Catal. 1980, 64, 18

12. Stoukides, M.; Vayenas, C.G. J. Catal. 1981, 69, 18

13. Sato, M.; Seo, J. J. Catal. 1972, 24, 224

14. Flank, W.H.; Beachell, H.C. J. Catal. 1967, 8, 316

15. Force, E.L.; Bell, A.T. J. Catal. 1976, 44, 175

16. Carberry, J.J.; Kuczynski, G.C.; Martinez, E. J. Catal. 1972, 26, 247

17. Stoukides, M.; Vayenas, C.G. J. Catal. 1981, in press

18. Pancharatnam, S., Huggins, R.A.; Mason, D.M. J. Electrochem. Soc. 1975, 122, 869.

19. Gur, T.M.; Huggins, R.A. J. Electrochem. Soc. 1979, 126, 1067

20. Vayenas, C.G.; Farr, R.D. Science 1980, 208, 593

21. Farr, R.D.; Vayenas, C.G. J. Electrochem. Soc. 1980, 127, (7), 1478

22. Vayenas, C.G.; Lee, B.; Michaels, J. J. Catal. 1980, 66, 36

23. Vayenas, C.G.; Georgakis, C.; Michaels, J.; Tormo, J. Catal. 1981, 67, 348

24. Holbrook, L.L.; Wise, H. J. Catal. 1975, 38, 294

25. Bockris, J. O'M.; Reddy, A.K.N. "Modern Electrochemistry", New York 1973; P. 1020-1036; 1190-1195

RECEIVED July 28, 1981.

Characterization of the Adsorbed Layer of a Silver Catalyst in the Oxidation of Ethylene from Its Transient Adsorption Behavior

M. KOBAYASHI

Kitami Institute of Technology, Department of Industrial Chemistry, 090 Kitami, Hokkaido, Japan

The adsorbed layer on a silver catalyst surface, during the course of ethylene oxidation at 91°C, was characterized by examining the transient adsorption behavior of CO_2. The analysis of the results obtained by transient response, thermal desorption and pulse techniques suggested the existence of three different adsorbed oxygen species: isolated monoatomic oxygen (O_s^i), adjacent monoatomic oxygen (O_s^a) and diatomic oxygen species (O_s^d), differing in their reactivities. Carbon dioxide and ethylene are competitively adsorbed on these three oxygen species and the strength of CO_2 adsorption on O_s^i is weaker than C_2H_4. CO_2 and C_2H_4 are not adsorbed on the surface reduced by H_2 but adsorbed on the surface reduced by C_2H_4 on which there is O_s^i. The C_2H_4 adsorbed on O_s^i produces no C_2H_4O and CO_2, the C_2H_4 on O_s^a produces an intermediate (In) in the complete oxidation and C_2H_4 on O_s^d produces C_2H_4O.

Desorption of reversibly and irreversibly adsorbed CO_2 is accelerated by the removal of adsorbed oxygen with H_2. The adsorption isotherm for CO_2 obtained on an oxidized surface reaches saturation for $Pco_2=0.15$ atm. The addition of $Pco_2=0.15$ atm into a reaction gas stream retarded the rates of C_2H_4O and CO_2 formation to zero in spite of a presence of adsorbed C_2H_4, indicating the complete blocking of CO_2 to the adsorbed oxygen and no reaction of the adsorbed C_2H_4 with gaseous O_2. The adsorption of CO_2 during reaction obeyed a Langmuir adsorption isotherm. The catalyst surface under steady state reaction is characterized as follows: appreciable parts of the silver surface larger than 0.4 is blocked by (In), irreversibly adsorbed CO_2 and adsorbed C_2H_4. It is sugg-

0097-6156/82/0178-0209$07.25/0
© 1982 American Chemical Society

ested that a very small fractional part of surface oxygen is available for the reaction at higher concentration of ethylene.

A large number of investigators have studied the
adsorption behavior of oxygen, carbon dioxide and
ethylene on silver catalysts. The chemisorption data
using a variety of techniques have suggested the following (1): oxygen is chemisorbed on silver surfaces
to form both monoatomic and diatomic oxygen species
based on a rough classification, ethylene is chemisorbed on silver ions (2) whereas no its adsorption
occurs on reduced silver (3) and carbon dioxide is
chemisorbed on previously oxidized surfaces (4,5).
Based on these experimental findings it is drawn that
adsorbed oxygen species should play an important role
to form adsorbed intermediates and adsorption sites
for reaction gas components. Therefore, to know the
detailed reaction mechanism of ethylene oxidation, it
is necessary to clarify a situation of the adsorbed
layer formed during the reaction, especially on the
adsorbed oxygen species available for the progress of
reaction.

Carbon dioxide adsorbs on adsorbed oxygen species,
not bare silver surfaces, in competition with other
reaction gas components and retards the overall reaction. One can conveniently use this nature to reveal
the situation of the adsorbed layer during the reaction.
For example, the adsorption isotherm of CO_2 obtained on
the surface used for reaction will present an efficient
information for the amounts of the adsorbed oxygen
species free from adsorbed intermediates, when the
intermediates are irreversibly adsorbed and stably
retained on the surface. And an examination for the
competitive adsorption of CO_2 and C_2H_4 on its surface
will propose an information for their adsorption strength and/or for a reactivity of the adsorbed ethylene to
react with the adsorbed oxygen or with gaseous oxygen.
In the present study, the transient response method (6,
7,8) has been efficiently used to follow the competitive
adsorption behavior between CO_2 and other reaction gas
components in ethylene oxidation, and thereby the adsorbed layer of a silver catalyst was characterized. A
low reaction temperature as 90°C and a large amount of
catalyst were used to follow more clearly the transient
states of adsorption and desorption of reaction components.

Experimental Method

The silver catalyst was prepared by reducing silver oxide. The silver oxide used was prepared by adding a solution of potasium hydroxide to an aqueous solution of silver nitrate. A small amount of 0.3% K_2SO_4 solution was added to the silver oxide powder as a promoter and, after mixing, was dried at 105°C for 24hr in a dark room. This was coated on α-Al_2O_3 of 20-42 mesh in the presence of a small amount of ethanol until the sample reached a size of 12-14 mesh. After the ethanol in the silver oxide powder had been completely vaporized in air at room temperature, the sample was reduced in a reactor with a flow of H_2 for 12 hr at 50°C and successively for 12 hr at 100°C. The composition of the catalyst so prepared was 206.0 g-Ag, 1.13 g-K_2SO_4/53.5 g-Al_2O_3. The BET surface area was 0.3 m^2/g-Ag. The constant activity of this catalyst was obtained by flowing the mixture of 5% C_2H_4, 20% O_2 and 75% He at 91°C for 48 hr.

The reactor used consists of a Pyrex glass tube containing 260.6 g catalyst and was immersed in an oil bath. The temperature of the catalyst bed remained constant within $\pm$ 0.1°C of the desired temperature. Oxygen (O_2 99.9%), nitrogen (N_2 99.9%), nitrous oxide (N_2O 99.99%), carbon dioxide (99.9%) and helium (He 99.999%) from commercial cylinders were purified through a dry ice-methanol trap to remove water vapor. For He the further purification to remove oxygen was conducted by passing through molecular sieves 5A which was cooled at the liquid nitrogen temperature. Ethylene (C_2H_4 99.9%) from a commercial cylinder was used without treatment.

The total flow rate of the gas was kept constant at 160($\pm$ 2) ml (NTP)/min and the composition of reaction mixture was varied by changing the concentration of helium as a diluent. The transient response to a change in the composition of He-N_2 mixture was completed within 15 sec. The intraparticle diffusion resistance was confirmed to be negligible by examining the rate data for catalysts of different sizes, 12-14 mesh and 20-42 mesh at 91°C. The external mass transport effect was also found to be negligible at 123°C by examining the rate data at constant W/F with various flow rates and catalyst amounts. The reaction conditions were chosen in such a way that the total conversion of ethylene did not exceed 0.08 in all experiments. In this condition, it was confirmed, by analysing the gas at five positions along the reactor length during the reaction, that the concentration of products (C_2H_4O and CO_2) increased linearly from the entrance of the catalyst bed to the exit.

Three gas chromatographes, each kept under different conditions, were simultaneously used to analyse all reaction gas components as continuously as possible. In the special case of a very rapid response with the period shorter than a few minutes, the same response was repeated a few times and the results obtained were superimposed in order to draw a response curve as continuously as possible.

Three flow systems with different gas composition were prepared so that the transient response experiments could be completed for three different gas mixtures within a few minutes. A more detailed description of transient response method used in this study can be found elsewhere ($\underline{6},\underline{7},\underline{8}$).

Experimental Results and Discussion

The response of the component Y in the outlet gas mixture to a step change in the concentration of X in the inlet gas stream is designated as X-Y response. The following symbols will be used:when X is increased, X(inc.,)-Y; when X is decreased, X(dec.,)-Y; when X is increased from nil, X(inc.,0)-Y; when decreased to nil X(dec.,0)-Y; when X is pulsed in the inlet gas stream, X(pulse)-Y response; when the temperature of the catalyst bed is raised linearly with respect to the elapsed time, T(linear)-Y response.

Transient Behavior of Oxygen

In order to analyze the adsorption behavior of carbon dioxide on silver it was necessary to understand the adsorption behavior of O_2 and its reactivity, because the adsorption of CO_2 strongly related to the adsorbed oxygen species as will be described later. For this reason, the following transient experiments were purformed.

Total Amount of Adsorbed Oxygen. The amount of adsorbed oxygen can easily be estimated from the graphical integration of the O_2-O_2 response curve. After the catalyst surface had completely been reduced by a pure hydrogen stream for about 5 hrs, the stream was switched over to a pure helium stream and then the O_2(inc.,0)-O_2, the O_2(dec.,0)-O_2 and again the O_2(inc.,0)-O_2 responses were followed successively. This procedure was repeated after changing the concentration of O_2. Fig.1 clearly shows a delay for the O_2(inc.,0)-O_2 response (Run1) at first time and shows an instantaneous response for both the successive O_2(dec.,0)-O_2

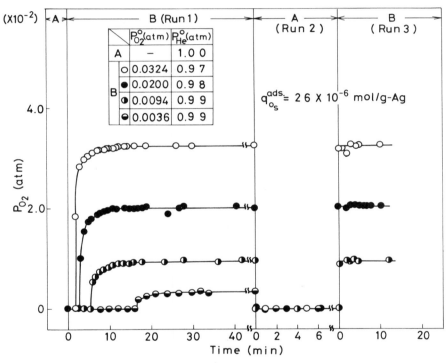

Figure 1. O_2–O_2 response at 91°C.

(Run 2) and O_2(inc.,0)-O_2 (Run 3) responses at second
time. These results indicated an irreversible adsorption of O_2 and no appreciable adsorption of oxygen on
the oxidized surface. The integrated amount of the
oxygen with Run 1 was estimated to be 2.6($\pm$0.1)x10^{-6}
mol/g-Ag (1.04($\pm$0.04)x10^{19} atoms/m^2) no dependence on
the concentration of O_2.

The adsorption strength of the adsorbed oxygen can
be determined by a thermal desorption technique. After
the catalyst surface had completely been oxidized in
the O_2(20%)-He mixture at 91°C, the reactor was flushed
with a pure He stream and temperature of catalyst bed
was then raised linearly at a rate of 4.3°C/min from
50 to 280°C. Thus the T(linear)-O_2 response was followed. Fig.2 shows that the desorption spectrum of oxygen
has a peak at 250°C. The integrated amount of desorbed
oxygen was estimated to be 2.5($\pm$0.1)x10^{-6} mol/g-Ag
(1.00($\pm$0.04)x10^{19} atoms/m^2), which is very close to the
value estimated from the O_2(inc.,0)-O_2 response. From
these results, it was apparent that all the oxygen
species adsorbed at 91°C were desorbed at around 250°C.
Scholten, Konvalinka and Beekman (9) calculated that
the total number of surface silver atoms was 1.31x10^{19}
atoms per m^2 based on Sandquist's data (10), and their
experiments on oxygen adsorption gave the number of
adsorbed oxygen atoms to be about 0.7x10^{19} at 50°C and
about 1.2x10^{19} atoms/m^2 at 100°C. Thus the total
amount of adsorbed oxygen obtained in this study fell
within Scholten et al's range and corresponds to 0.8 of
the surface coverage expected for a monoatomic form of
adsorbed oxygen.

The surface reduced by H_2 was active for the
decomposition of N_2O at 91°C. After the catalyst had
been reduced by the H_2 stream for 5 hrs at 91°C, the
reactor was flushed by the He stream and then the N_2O
(inc.,0)-N_2 response (Run 1) was followed. Run 1 in
Fig.3 showed a typical overshoot type response of N_2
with an instantaneous maximum. This instantaneous mode
indicated the direct decomposition of gaseous N_2O on
active sites: this is also suggested by the N_2O(dec.,0)
-N_2O response (Run 2) in Fig.3 which exhibited no delay
indicating no adsorption of N_2O. The graphical integration of the N_2O(inc.,0)-N_2 response curve corresponded to the amount of adsorbed oxygen atoms and was
estimated to be 2.2x10^{-6} mol/g-Ag (0.90($\pm$0.08)x10^{19}
atoms/m^2). This is close to the value 0.80x10^{19} which
is roughly estimated from the figure given by Scholten
et al. The difference between the amount of adsorbed
oxygen found oxygen adsorption and as compared to N_2O
decomposition may be attributed to the presence of

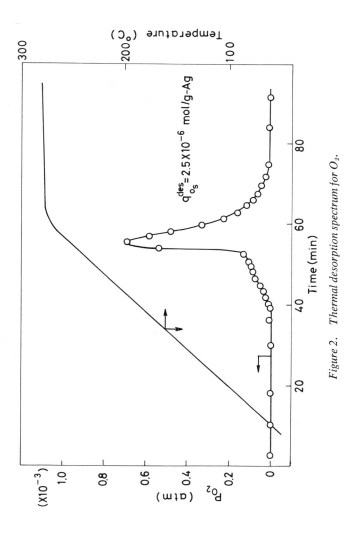

Figure 2. Thermal desorption spectrum for O_2.

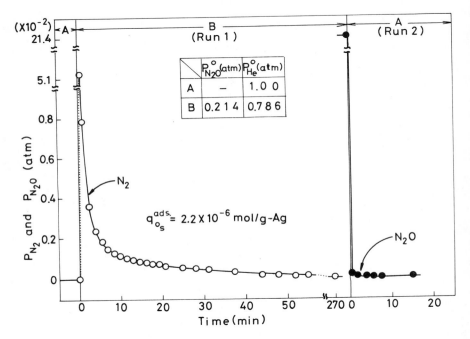

Figure 3. $N_2O(inc. O)$–N_2 *and* $N_2O(dec. O)$–N_2O *responses at* $91°C$.

adsorbed oxygen in a diatomic form. The amount of
diatomic oxgen in the case of the O_2 adsorption was
estimated to be 0.4×10^{19} molecules/m^2, disregarding
the recombination of atomic oxygen formed during the
N_2O decomposition.

Reactivity of Adsorbed Oxygen. The existence of
two types of adsorbed oxygen on silver, diatomic and
monoatomic oxygen, has frequently been claimed ([1]). In
the present study the catalytic activity of the two
types of adsorbed oxygen was also investigated again
using the pulse technique. Before use the pulse technique one should examine the desorption behavior of
ethylene oxide because the pulse spectrum of the ethylene oxide formed by a reaction of ethylene with adsorbed oxygen species is strongly affected by it.

After the catalyst had been oxidized in an O_2-He
stream for 5 hrs, the stream was switched over to
either a C_2H_4-O_2-He mixture (Run 1) or a C_2H_4-He mixture (Run 2) and the C_2H_4, O_2(inc.,0)-C_2H_4O and -CO_2
and the C_2H_4(inc.,0)-C_2H_4O and -CO_2 responses were
followed separately. All the responses in the two
Runs were of the overshoot type as shown in Fig.4.
During the initial stage of the responses, C_2H_4O and
CO_2 in two Runs reached the same maximum points irrespective of whether oxygen existed in the gas phase.
This result strongly indicated that adsorbed oxygen not
gaseous oxygen was responsible for the production of
C_2H_4O and CO_2. Furthermore, the response of C_2H_4O was
of the overshoot type with an instantaneous maximum.
Such behavior probably resulted from the rapid desorption of the C_2H_4O produced. This idea can also be
supported by the following results.

The height at the maximum point of C_2H_4,O_2(inc.,0)
-C_2H_4O response in Fig.4 corresponds to the reaction
rate ($r^{max}_{C_2H_4O}$) of adsorbed oxygen with C_2H_4. The $r^{max}_{C_2H_4O}$
was plotted as a function of Pc_2H_4 and a good straight
line obtained at temperatures ranging from 80 to 91°C
as shown in Fig.5. The Arrhenius plot of the apparent
rate constant estimated from these straight lines was
58.6 kJ as shown in Fig.6. These results strongly
suggest that C_2H_4O is formed by the reaction between
gaseous C_2H_4 and adsorbed oxygen, an Eley-Rideal type
mechanism, and that the C_2H_4O formed desorbs rapidly.

In such a case, pulse techniques can be conveniently used to examine the reactivity of the adsorbed species as follows. After the catalyst had been reduced
by a pure H_2 stream at 160°C for 2 hrs, the catalyst
bed was flushed by He and then the sream changed to an
C_2H_4(12.8%)-He stream. O_2 and N_2O gases were separa-

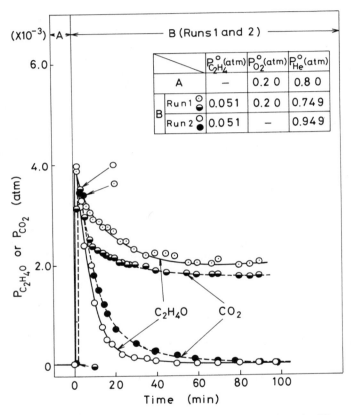

Figure 4.   Comparison of Run 1, the C_2H_4(inc. O)–C_2H_4O and –CO_2 responses and Run 2, the C_2H_4(inc. O)– and O_2(dec. O)–C_2H_4O and –CO_2 responses at 91°C.

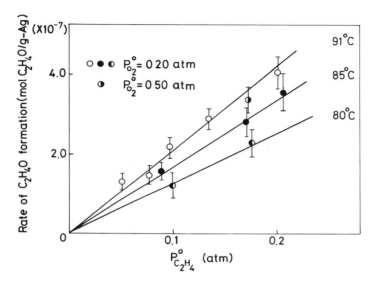

Figure 5.   The rate of C_2H_4O formation at the maximum point of the C_2H_4(inc. O)–
C_2H_4O response.

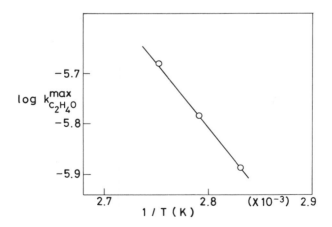

Figure 6.   Arrhenius plot for $k_{C_2H_4O}{}^{max}$ at $P_{O_2}{}^\circ = 0.2 - 0.5$ atm.

tely pulsed into the inlet stream of this mixture in various pulse sizes. The O_2(pulse)-C_2H_4O and -CO_2 responses and the N_2O(pulse)-C_2H_4O and -CO_2 responses were separately followed. The same measurements were repeated three times for each pulse and typical data obtained were presented in Figs.7 and 8 by Run 1.

Comparison shows that, in the case of the O_2-pulse, co-production of CO_2 and C_2H_4O was observed whereas with N_2O, CO_2 was mainly formed. One can roughly estimate the total amount of each oxygen species, taken up by the catalyst surface from each pulse, by the application of simple stoichiometry to the amount of the produced CO_2 and C_2H_4O and an intermediate (In) in the complete oxidation. Here the existence of the intermediate (In) was demonstrated in our previous papers (11,12). The amount of (In) is obtained from the amount of CO_2 formed by the decomposition of it in an O_2-He stream (see Run 2 in Figs.7 and 8). In addition it is assumed that (In) keeps the carbon-carbon bond and retains four atomic oxygen (11). Fig. 9 indicates that the adsorbed oxygen species from the O_2 pulses produced C_2H_4O more than hundred times faster than that from N_2O pulse. This is consistent with Herzog's results (13) which demonstrated very low selectivity to C_2H_4O from C_2H_4-N_2O under steady state conditions at temperatures above 210°C. Furthermore, at 91°C, the oxygen species formed with the N_2O pulses produced only (In) with no production of CO_2 or C_2H_4O, and the formed (In) was decomposed to the equal amounts of CO_2 and H_2O in an O_2-He stream whereas no its decomposition occured in a N_2O-He stream (13).

From the above results, one can conclude the existence of the two types of adsorbed oxygen species which may be speculated as monoatomic and diatomic oxygen species. The monoatomic oxygen species can produce (In) and the diatomic oxygen species can produce C_2H_4O and decompose (In).

Adsorption Behavior of C_2H_4

Adsorption behavior of C_2H_4 was examined on three different surfaces: those reduced with H_2, those reduced with C_2H_4 and those previously used in reaction. On the surface reduced by H_2 the C_2H_4-C_2H_4 response simply was an instantaneous response as shown in Fig. 10. This showed that C_2H_4 did not adsorb on bare silver surface. In fact, no production of CO_2 or desorption of C_2H_4 was observed even when the catalyst surface was exposed to the O_2-He mixture following the C_2H_4-C_2H_4 response experiment.

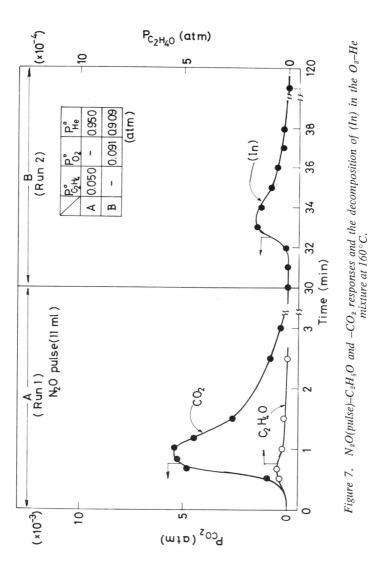

Figure 7. N₂O(pulse)–C₂H₄O and –CO₂ responses and the decomposition of (In) in the O₂–He mixture at 160°C.

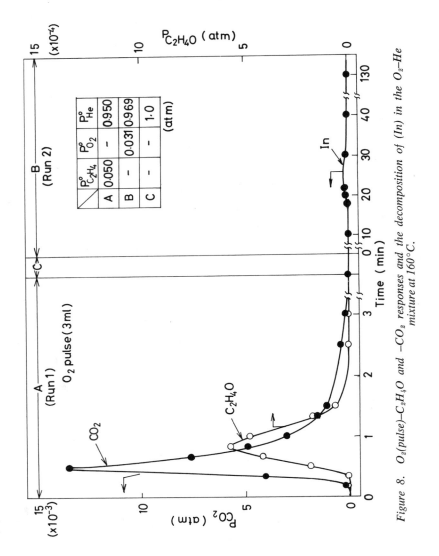

Figure 8. $O_2(pulse)$–C_2H_4O and $-CO_2$ responses and the decomposition of (In) in the O_2–He mixture at 160°C.

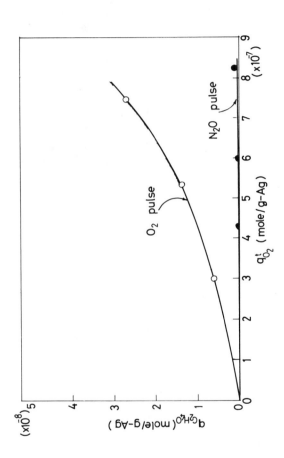

Figure 9. Comparison of the amounts of C_2H_4O formed by the O_2 and N_2O pulses at 160°C.

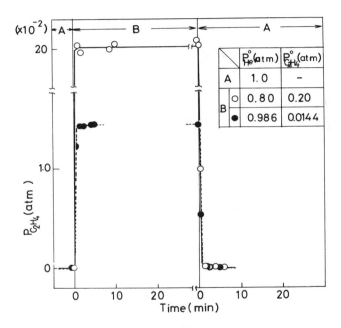

Figure 10. C_2H_4–C_2H_4 response on the reduced surface with H_2 at 91°C.

The adsorption behavior of C_2H_4 on a surface which had been reduced by C_2H_4-He stream for 24 hrs until no CO_2 and C_2H_4O was rather different. The C_2H_4(dec.,0)-C_2H_4 response clearly showed a delay indicating the existene of an appreciable amount of adsorbed C_2H_4 on the surface. The integrated amount of adsorbed C_2H_4 plotted as a function of the pressure of C_2H_4 was as shown in Fig.11 by curve I. This adsorption isotherm gave the amount of C_2H_4 at saturation to be $3.8(\pm0.4)$ $x10^{-7}$ mol/g-Ag $(7.6x10^{17}/m^2)$. Since C_2H_4 did not adsorb on bare silver, adsorbed oxygen species must be the active sites for the adsorption of C_2H_4. Furthermore this adsorbed oxygen could produce neither CO_2 nor C_2H_4O, being only active sites for C_2H_4 adsorption. The diatomic oxygen species can not be these active sites, because they form C_2H_4O immediately by the reaction. Thus monoatomic adsorbed oxygen species are believed to be the sites. As described previously, since the monoatomic oxygen species can also form the intermediate (In) in the complete oxidation of ethylene, it may reasonably be considered that monoatomic oxygen species are of two types: one is those consisting of more than two atoms and which can produce (In) and those present as isolated monoatomic oxygen species and which are the active sites for C_2H_4 adsorption. These isolated oxygen species may be able to migrate on surface to make adjacent oxygen species or diatomic oxygen species. It is these isolated monoatomic oxygen which exist on the reduced surface with C_2H_4. Using this idea one can also explain the following experimental results.

The amount of C_2H_4 adsorbed during reaction can easily be estimated from the graphical integration of the C_2H_4,O_2(dec.,0)-C_2H_4 response curves obtained at steady state. The result is presented in Fig.11 by Curve I and shows a linear adsorption isotherm. It was also determined that this isotherm was not affected by the presence of oxygen in the gas phase during the course of the C_2H_4 desorption. This strongly indicates that the isolated monoatomic oxygen species also exist on the surface during reaction and that C_2H_4 adsorbed on these isolated species does not produce any products. i.e. the adsorbed C_2H_4 is inactive for the reaction. Furthermore, when the surface reduced with C_2H_4 was kept in a pure He stream for several hours, and the surface was exposed again to the C_2H_4-He stream, a small amount of C_2H_4O was detected in the effluent gas stream. This suggests that isolated monoatomic oxygen migrates very slowly on surface to form the diatomic oxygen species.

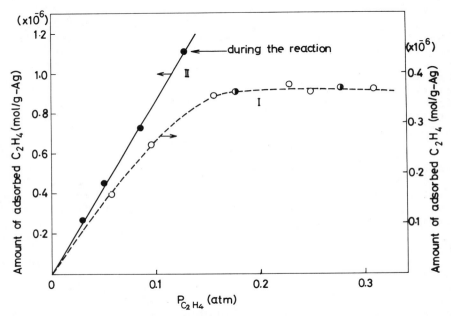

Figure 11. Adsorption isotherms for C_2H_4 on the surface reduced with C_2H_4 and on the surface used for the reaction at $91^\circ C$. Key: ○, in He; and ◑, in CO_2–He.

Transient Behavior of CO_2

Adsorption Behavior of CO_2. The adsorption behavior of CO_2 on silver is greatly influenced by whether
the surface was reduced or oxidized. In the present
study four different surfaces were studied: a surface
reduced by H_2, a surface reduced by C_2H_4, an oxidized
surface and a surface used for reaction. The CO_2(inc.,
0)-CO_2 and CO_2(dec.,0)-CO_2 responses for a surface
reduced by H_2 exhibited no delay, indicating no adsorption of CO_2 on the bare silver surface in agreement
with the works of Drake and Benton (4) and of Czanderna
(5).
      When the CO_2(inc.,0)-CO_2 and CO_2(dec.,0)-CO_2
responses were followed in a He stream on a surface
reduced by C_2H_4, the response showed a delay indicating
a reversible adsorption of CO_2. The integrated amount
of the adsorbed CO_2 was estimated to be 4.0×10^{-7} mol/
g-Ag ($8 \times 10^{17}/m^2$) independent of concentration of CO_2
as shown by its adsorption isotherm (Curve II in Fig.
12). This is close to the saturation amount of adsorbed C_2H_4 on the same surface (compare to Curve I in
Fig.11). This good agreement indicates that isolated
monoatomic oxygen species are also the active sites
for the adsorption of CO_2. In contrast, CO_2-CO_2 responses conducted in a C_2H_4(16%)-He stream instead of
in the He stream were instantaneous indicating no adsorption of CO_2 (see Curve IV in Fig.12). This is a
consequence of competitive adsorption between C_2H_4 and
CO_2 on the same isolated monoatomic oxygen species with
adsorption of C_2H_4 being stronger than CO_2.
      After the catalyst had completely been oxidized in
the O_2(20%)-He stream for 24 hrs, the CO_2-CO_2 response
was followed in a He stream. The amount of adsorbed
CO_2 estimated from the CO_2(inc.,0)-CO_2 response was in
good agreement with the amount of desorbed CO_2 estimated from the CO_2(dec.,0)-CO_2 response, except the
amount of adsorbed CO_2 estimated from the first CO_2
(inc.,0)-CO_2 response was more than for subsequent
responses. The difference was estimated to be $6(\pm1) \times$
10^{-7} mol/g-Ag independent on the concentration of CO_2.
This disagreement is considered due to the existence
of some irreversible adsorption of CO_2. The existence
of this irreversible adsorption was confirmed by the
following experiment.
      After several CO_2(inc.,0)-CO_2 responses had been
completed, the CO_2-He mixture was changed to pure H_2
and the CO_2(dec.,0)-CO_2 response was followed. As seen
in Fig.13 these was large amount of CO_2 desorption
during the initial stage of the response followed by a

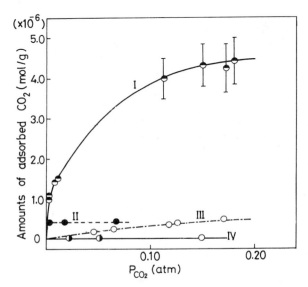

Figure 12.   Adsorption isotherms at 91°C for CO_2 on the differently treated catalyst surfaces.   Key: I, ◓, in He (oxidized surface); II, ●, in He (surface reduced by C_2H_4); III, ○, in He (surface used for the reaction); and IV, ◑, in C_2H_4–He (surface reduced by C_2H_4) and ○, in He (surface reduced by H_2).

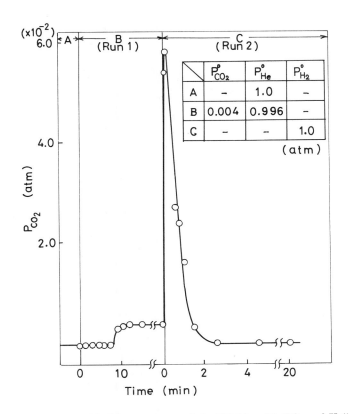

Figure 13.   CO₂(inc. O)–CO₂ response and the CO₂(dec. O)–CO₂ and H₂(inc. O)– CO₂ response at 91°C.

steep decline to zero within a few minutes. The mode
of the response is typical of false start responses and
may be attributed to an acceleration of the desorption
rate of CO_2 caused by the rapid removal of adsorbed
oxygen by H_2 reduction. The integrated amount of the
desorbed CO_2 in Run 2 is larger than that estimated
from the $CO_2(dec.,0)-CO_2$ response in a He stream and
the difference is estimated to be 7 ($\pm$1)x10^{-7} mol/g-Ag
(1.4x10^{18}/m^2). This corresponds to the amount of the
irreversibly adsorbed CO_2 estimated previously. One
can conclude that the desorption rates of both reversibly and irreversibly adsorbed CO_2 are greatly accelerated by the reduction of surface with H_2.

The adsorption isotherm for reversibly adsorbed
CO_2 estimated from the CO_2-CO_2 response is presented in
Fig.12 by Curve I. From Curve I the saturated amount
of CO_2 is estimated to be 4.3($\pm$0.5)x10^{-6} mol/g-Ag
(8.6x10^{18}/m^2). This amount plus the amount of the
irreversibly adsorbed CO_2 is about 5x10^{-6} mol/g-Ag
which is very close to the amount of adsorbed oxygen
atoms estimated from the adsorption of O_2. This good
agreement strongly indicates that all adsorbed oxygen
species are blocked by the adsorbed CO_2 at a partial
pressure of CO_2 higher than P_{CO_2}= 0.15 atm: probably
one carbon dioxide molecule adsorbs on one adsorbed
oxygen species in which both diatomic and monatomic
oxygen are included.

Since the reaction rate is strongly retarded by
the adsorption of CO_2, it is difficult to measure the
adsorption behavior of CO_2 during the reaction. However one can roughly estimate the amount of CO_2 adsorbed on the surface which has been employed for reaction,
by using the following experiments. After the reaction
had reached a steady state, the C_2H_4, $O_2(dec.,0)-CO_2$
response (Run 1) was followed. Run 1 in Fig.14 shows
a very slow out flow of CO_2 for an extraordinarily
long period of time: this may be attributed to slow
surface migration of adsorbed oxygen to react with (In).
In a separate experiment, the same $C_2H_4,O_2(dec.,0)-CO_2$
response was followed for 120 min and a CO_2-CO_2 response measured. The result obtained is presented in
Fig.14 by Run 2. This CO_2-CO_2 response should be
considered as the response which was superposed with
the pseudo steady state level of the $C_2H_4,O_2(dec.,0)$
-CO_2 response (Run 1). The amount of adsorbed CO_2 (
q_{CO_2}) can be estimated from the difference between both
the $CO_2(dec.,0)-CO_2$ and the C_2H_4, $O_2(dec.,0)-CO_2$ response curves as schematically shown in Run 2 in Fig.14
by the shaded area. The adsorption isotherm for CO_2
thus obtained is presented in Fig.12 by Curve III. This

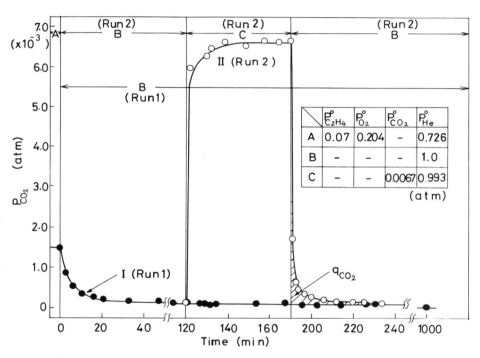

Figure 14. CO_2–CO_2 response at 91 °C on the surface used for the reaction.

curve obeys Langmuir adsorption isotherm with a saturation amount of adsorbed CO_2 of about 0.9×10^{-6} mol/
g-Ag (1.8×10^{18}/m^2).

Let us consider the composition of the adsorbed
layer of catalyst during reaction. The amount of (In)
at a steady state is given by the amount of CO_2 during
its decomposition in a O_2-He stream and is estimated
to range from 1.4×10^{-6} to 2.5×10^{-6} mol CO_2/g-Ag depending on the concentration of ethylene. Assuming that
one carbon dioxide molecule formed from (In) arose
from one adsorbed atomic oxygen, the above amount
exactly corresponded to the amount of the adsorbed
atomic oxygen which had been blocked by (In). On the
other hand, C_2H_4 and CO_2(rev.) were also reversibly
adsorbed on adsorbed oxygen species to block the reaction and the amounts of them, by using each adsorption
isotherm (see Cuve II in Fig.11 and Cuve III in Fig.
12), could be estimated to be 0.2-1.2×10^{-6} and 0.5-3
$\times 10^{-7}$ mol/g-Ag, respectively, the ranges being the
result of a dependence on the concentration of C_2H_4.
In addition, one has to take into account the existence
of the adsorbed oxygen blocked by the irreversibly
adsorbed CO_2(irr.) which is estimated to be 6×10^{-7} mol/
g-Ag. Finally, assuming that one CO_2 or C_2H_4 molecule
occupies one adsorbed monoatomic oxygen species, the
total amount of adsorbed oxygen blocked by the adsorption of (In), C_2H_4, CO_2(irr.) and CO_2(rev.) are roughly
estimated to range from 1×10^{-6} to 2×10^{-6} mol/g-Ag
depending on the concentration of ethylene. The upper
values are equal to the total amount of adsorbed oxygen
estimated from the decomposition of N_2O. This agreement strongly suggests that that appreciable part of
the ctalyst surface ($\theta = 0.8$) is oxidized during the
reaction and the fractional part of those surface
oxygen higher than 0.4 is blocked by the adsorption of
(In), C_2H_4 and CO_2(irr.). When the high concentration
of C_2H_4 is used, very small fractional part of the
surface oxygen is available for the reaction. This is
also consistent with the steady state rate data in
which the rates of C_2H_4O and CO_2 formation are gradually lowered with increasing the partial pressure of
ethylene.

Behavior of CO_2 during the Reaction. The effect
of CO_2 on both the rates of C_2H_4 formation and complete
oxidation was examined by adding CO_2 into the reaction
gas stream at the steady state of reaction. The rate
of complete oxidation was measured by the rate of H_2O
formation. Fig.15 clearly indicated that both rates
were strongly retarded by the addition of CO_2: when

$P^{\circ}_{C_2H_4}$ was 0.13 atm, C_2H_4O formation ceased when P_{CO_2} 0.13 atm and complete oxidation ceased when $P_{CO_2} \geq 0.11$ atm. with $P^{\circ}_{C_2H_4} = 0.27$ atm   the same was true when $P^{\circ}_{CO_2} = 0.18$ and 0.12 atm respectively. Thus the degree of the retardation depended on the concentration of C_2H_4. This is probably caused by the effect of adsorbed C_2H_4 which inhibits the adsorption of CO_2 onto the adjacent adsorbed oxygen, taking into account the previous fact that the strength of C_2H_4 adsorption was larger than that of CO_2.

Noting the selectivity to C_2H_4O in Fig.16, it gradually increased with the concentration of CO_2. The selectivity estimates involve much experimental error because of the difficulty of the accurate detection of the small amount of water produced. However the increase may be real and can be accounted for explained in terms of the various adsorbed oxygen species: a fractional part of the adjacent adsorbed oxygen species is decreased with increasing the concentration of CO_2 by the blocking effect of adsorbed CO_2 onto the adjacent oxygen. A fractional part of the isolated adsorbed oxygen is thus increased. The isolated diatomic oxygen can produce C_2H_4O, but the isolated monoatomic oxygen can not produce CO_2. This is the reason why the selectivity to C_2H_4O increased with increasing the adsorption of CO_2. This idea is also available to explain the following competitive adsorption behavior of CO_2 and C_2H_4.

After the oxidized surface had completely been blocked by the adsorption of CO_2 from a stream of CO_2 (18%)-He mixture (Run 1), a C_2H_4O-O_2-CO_2-He stream was introduced into the reactor in a stepwise fashion (Run 2) and C_2H_4,O_2(inc.,0)-C_2H_4O and -H_2O responses were followed. The results are presented in Fig.16. Run 2 in Fig.16 exhibited no production of C_2H_4O and H_2O, indicating complete blocking by CO_2 of both the diatomic and monoatomic oxygen species. Then the C_2H_4-O_2-CO_2-He stream was switched over to either the O_2-He stream (Run 3) or C_2H_4-O_2-He stream (Run 4) and the responses of CO_2, C_2H_4O and C_2H_4 were followed. Run 3 in Fig.16 clearly shows the desorption of C_2H_4 and no production of H_2O (this is not shown in the figure). From these results one can conclude that adsorbed C_2H_4 can not react with either gaseous oxygen or adsorbed oxygen to produce C_2H_4O and (In). This means that the adsorbed C_2H_4 just sits on the adsorbed oxygen species which are isolated from the neighboring adsorbed oxygen by the blocking effect of the adsorbed CO_2.

On the other hand, Run 4 in Fig.16 shows the slow increase for the rate of C_2H_4O formation and the

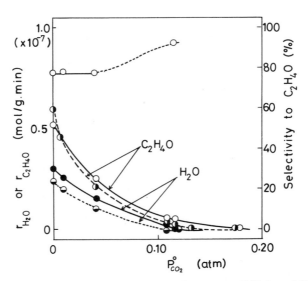

Figure 15. The retardation effect of CO_2 on the rates of C_2H_4O and H_2O formation at 91°C. Key: $\bigcirc$, $\bullet$, $P°_{C_2H_4} = 0.27$ atm, $P°_{O_2} = 0.21$ atm; and $\obullet$, $\obullet$, $P°_{C_2H_4} = 0.13$ atm, $P°_{O_2} = 0.21$ atm.

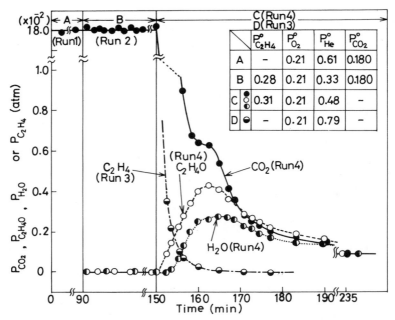

Figure 16. The effect of CO_2 adsorption on the C_2H_4(dec. O)–C_2H_4 response and on the C_2H_4(inc. O)–C_2H_4O and –H_2O responses at 91°C.

induction period for the H_2O formation at the initial
stage of the response, in contrast to the C_2H_4(inc.,0)
$-C_2H_4$ and $-CO_2$ responses in Fig.4.   These slow responses
are caused by the slow desorption of CO_2 from the
adsorbed oxygen species.   Here the reason for the induction period of H_2O may reasonably be attributed to the
evidence that CO_2 and H_2O are formed via the stable
intermediate (In).

The C_2H_4, CO_2(inc.,0)$-C_2H_4O$ and $-H_2O$ responses
were followed on the oxidized surface without previously treating the catalyst with the CO_2-O_2-He stream.
The results obtained at two different conentrations of
CO_2 are presented in Fig.17 as Runs 1($P°co_2=0.117$ atm)
and 2 ($P°co_2=0.18$ atm).   Appreciable amounts of C_2H_4O
and H_2O are produced at the initial stage of the responses in contrast to Run 2 in Fig.16.   The both
amounts strongly depend on the concentration of CO_2
introduced and decline steeply to near zero within 7
min.   This is resulted from the competitive attacking
of CO_2 and C_2H_4 onto the same adsorbed oxygen species.
The rate of CO_2 adsorption is not fast enough for
instantaneously blocking the adsorbed oxygen species,
faster than which react with C_2H_4.

After Runs 1 and 2 in Fig.17 had been followed for
2 hrs, the two $C_2H_4-O_2-CO_2-He$ mixtures were separately
replaced by the O_2-He mixture (Runs 3 and 4) and the
C_2H_4, CO_2(dec.,0)$-CO_2$ and $-H_2O$ responses were followed.
Runs 3 and 4 in fig.17 clearly indicated the production
of H_2O with progressive desorption of CO_2 from the
adsorbed oxygen species.   The total amount of the
produced H_2O strongly depends on the concentration of
CO_2 in Runs 1 and 2.   These results can reasonably be
explained as follows: the intermediate (In), which had
been formed during the course of Runs 1 and 2 and
retained on the surface without reacting with gaseous
O_2, is decomposed into H_2O and CO_2 by the reaction with
the adjacent adsorbed oxygen species which has been
free from the adsorbed CO_2.   The higher the concentration of CO_2 in Runs 1 and 2 in Fig.17, the smaller
amount of (In) is formed,because the degree of the
blocking effect to the adjacent adsorbed monoatomic
oxygen depends on the concentration of CO_2.

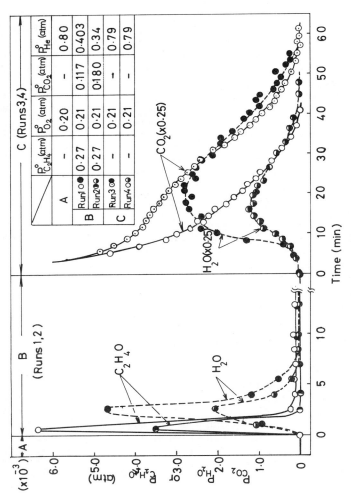

Figure 17.   The effect of CO₂ adsorption on the C₂H₄(inc. O)–C₂H₄O and H₂O responses and on the (In) formation at 91°C.

Acknowledgment

Thanks are due to Mr.Yamamoto for carrying out some of the experiments.

Literature Cited

1) Kilty,P.A. and Sachtler,W.M., Catalysis Rev.,1974, 10, 1.
2) Force,E.L. and Bell,A.T., J.Catal., 1975, 38, 440.
3) Gerei,S.V., Kalyavenko,K.M., and Rubanik,M.Ya, Ukr.Khim.Zh., 1965, 31(2), 166.
4) Drake,L.C., and Benton,A.F., J.A.C.S.,1934, 56, 506.
5) Czanderna,A.W., J.Colloid.Interface Sci., 1966, 22, 482.
6) Kobayashi,M. and Kobayashi,H., J.Catal., 1972, 27, 100, 108, 114; Catal.Reviews, 1975, 10, 139.
7) Kobayashi,M., Doctral thesis, Hokkaido Univ. 1975.
8) Kobayashi,M., Preprint, 5th Canadian Symposium on Catalysis, 1977, 202.
9) Scholten,J.J.F., Konvalinka,J.A. and Beekman,F.W., J.Catal., 1973, 28, 209.
10) Sundquist,B.E., Acta.Met., 1964, 12, 67.
11) Kobayashi,M., Yamamoto,M. and Kobayashi,H.,Proceeding.6th Intern.Cong.Catal., 1976, A24.
12) Kobayashi,M. and Kobayashi,H., J.C.S.,Chem.Comm., 1976, 103; ibd., 1977, 71.

RECEIVED July 28, 1981.

Development of a Pulse Reactor with On-Line MS Analysis to Study the Oxidation of Methanol

C. J. MACHIELS

E. I. du Pont de Nemours & Co., Inc., Central Research and Development Department, Experimental Station, Wilmington, DE 19898

Whereas steady state kinetic experiments are usually more easy to perform and certainly more easy to analyze, information is normally only obtained pertaining to the rate limiting step in the reaction sequence. Experiments of a dynamic type, such as pulse experiments or other transients, can give additional information about reaction mechanisms that is often difficult to obtain by other techniques. Additional advantages of pulse experiments are that they often take less time and require just small amounts of materials, both catalyst and reactants. The major problems of the pulse and other transient techniques seem to be related to the lack of good and rapid transient analytical methods and the fact that data analysis is not straightforward. Many good reviews are available concerning the theory and applications of pulse and other transient experiments (1,2,3).

In a typical pulse experiment, a pulse of known size, shape and composition is introduced to a reactor, preferably one with a simple flow pattern, either plug flow or well mixed. The response to the perturbation is then measured behind the reactor. A thermal conductivity detector can be used to compare the shape of the peaks before and after the reactor. This is usually done in the case of non-reacting systems, and moment analysis of the response curve can give information on diffusivities, mass transfer coefficients and adsorption constants. The typical pulse experiment in a reacting system traditionally uses GC analysis by leading the effluent from the reactor directly into a gas chromatographic column. This method yields conversions and selectivities for the total pulse, the time coordinate is lost. It is desirable to be able to continuously monitor all the reactants and products so that a time resolved response curve is obtained for the separate components in the pulse. The sequence in which the products appear and the shape of each individual peak can, in principle, give information on the reaction sequence and on adsorption properties that cannot easily be obtained in any other way. Infrared spectroscopy has been used with good results, but the number of systems suitable is limited and at best only a small number of components can be analyzed for, Figure 1. The

0097-6156/82/0178-0239$05.00/0

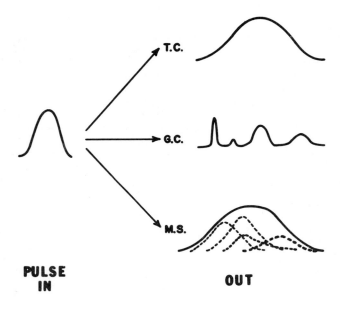

Figure 1.   Analytical methods for pulse experiments.

most general method is a mass spectrometer with continuous inlet system, and several reports have appeared recently describing such systems (4-8). Many of these systems lacked sufficient speed for pulse experiments, often an experiment had to be repeated several times, looking at a different component each time.

Methanol Oxidation

Initial tests using the pulse reactor described in this paper have been done on the selective oxidation of methanol to formaldehyde using molybdate catalysts.

A commercial catalyst from Harshaw was used, a 3:1 mixture of molybdenum trioxide and ferric molybdate, as well as the two separate phases. Kinetic experiments were done previously in a differential reactor with external recycle using these same catalysts as well as several other preparations of molybdenum trioxide, including supported samples. The steady state kinetic experiments were done in the temperature range 180-300 C, and besides formaldehyde, the following products were observed, dimethylether, dimethoxymethane, methylformate, and carbon-monoxide. Usually very little carbon dioxide was obtained, and under certain conditions, hydrogen and methane can be produced. The steady state experiments showed that the two separate phases and the mixture are not very different in activity, give approximately the same product distributions, and have similar kinetic parameters. The reaction is about .5 order in methanol, nearly zero order in oxygen, and has an apparent activation energy of 18-20 kcal/mol. These kinetic parameters are similar to those previously reported (9,10), but often ferric molybdate was regarded to be the major catalytically active phase, with the excess molybdenum trioxide serving for mechanical properties and increased surface area (10,11,12).

At least four different steps have been reported to be the rate determining step in this reaction, ranging from the adsorption of methanol to the desorption of the products and the reoxidation of the surface. Studies in the recycle reactor using deuterium labeled methanol showed that over a limited temperature range, the rate limiting step on the commercial catalyst is the breaking of a carbon-hydrogen bond (13). The ratios of rate constants in Table I show that a deuterium in the hydroxyl group has little effect on activity; selectivities are not affected. A primary kinetic isotope effect is observed, however, when completely deuterated methanol is used, and the product distribution also changes significantly; much more dimethylether is formed and less formaldehyde. For the formation of the ether, no carbon-hydrogen bonds need to be broken.

Table I

Kinetic isotope effect on $Fe_2(MoO_4)_3/MoO_3$

	$\dfrac{k_{CH_3OD}}{k_{CH_3OH}}$	$\dfrac{k_{CD_3OD}}{k_{CH_3OH}}$
220	0.93	–
230	0.92	0.26
260	0.94	0.37

Experimental

A flow schematic of the pulse reactor is shown in Figure 2. The flexible design allows dynamic experiments of various types. Helium carrier gas flows through the two channels of a Gow-Mac thermal conductivity cell before and after passing through the reactor. Pulses of methanol in argon or an argon-oxygen mixture can be introduced using a six-port Valco valve with a 1 cc sample loop. Pulses of methanol can be alternated with oxygen pulses. A four-port Valco valve allows application of step changes. For this purpose, the helium stream can be replaced by another methanol-containing stream. This set up can also be used for pulses of isotopically labeled species onto a steady state of unlabeled materials. Methanol streams are obtained using a saturator method and concentrations are determined by gas chromatographic analysis. Back pressure regulators in the reactor stream and the sample gas stream maintain an equal pressure of 1-2 psig in both streams. Pulses can be dispersed in a column before the reactor if desired.

The reactor itself Figure 3 is made of Pyrex, using 1mm capillary tubing wherever possible. The catalyst (0.2-0.5 g) is on a fritted disk, 14 mm in diameter. The catalyst bed is several mm high. A continuous sample to the mass spectrometer is taken from just above the catalyst bed. A short section of 1/16" stainless steel tubing is coupled to a 1-m stainless steel capillary, 0.15 mm i.d./2 mm o.d. Part of the reactor effluent is pumped through this capillary and through the inlet valve of the mass spectrometer. If this valve is opened, a small portion of the gas leaks through an orifice directly into the ion formation region of the ion source. The pressure reduction in two stages from atmospheric to approximately 1 mbar in the valve to 10^{-6}-10^{-5} mbar in the spectrometer ensures viscous flow in the capillary and minimizes mass discrimination effects. Good results were also obtained when the stainless capillary was replaced by a section of fused silica capillary.

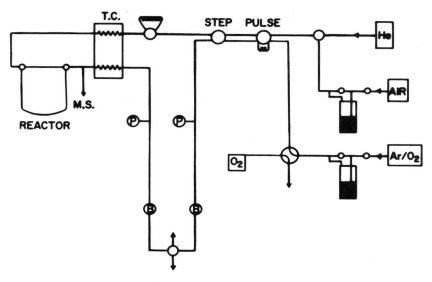

Figure 2.   Flow diagram of reactor for transient experiments.

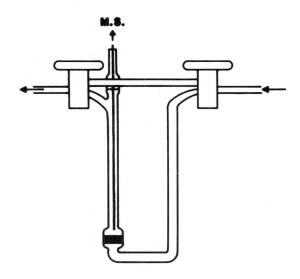

Figure 3.   Pyrex pulse reactor.

The quadrupole mass spectrometer used for analysis is a QMG
511 by Balzers. This system comes as a compact, mobile package,
uses turbomolecular pumping, and is interfaced to a DEC PDP-11/34
minicomputer. A limited software package allows full computer
control of all the spectrometer functions and various methods of
data acquisition. A gas tight axial ion source is used for high
ionization efficiency, and a 90° off axis electron multiplier
improves the signal to noise ratio. A data integrator allows for
maximum speed and stability of data acquisition; ion signals are
collected at 17 µs/point, and each 64 data points are averaged in
hardware in the integrator before transferring to the interface.
Peaks are symmetrical and have flattened tops, typically the
change of intensity at the top of a peak is not more than 5% over
1/4 of a mass unit. This is an important characteristic, because
with the rapid scanning during a pulse, there is no opportunity to
scan the complete mass range and determine the exact location of
each peak in each scan. Instead, we have to jump from the top of
one peak to the next and so on in a cyclical fashion. Flattened
tops ensure that slight shifts in peak location do not cause large
errors in this process. Figure 4 shows a flow diagram for
sample analysis. Raw intensity data are stored in core memory and
then transferred to disk after the experiment. They can be used
for plotting and further analysis. Before an experiment, the
exact location of the peaks of interest is determined and if
necessary a background spectrum is obtained. After a pulse is
taken, scanning is not started until triggered by mass 40 (argon).

To test the reactor and analysis system, pulses of methanol,
singly, and completely deuterated methanol were led over the
commercial $Fe_2(MoO_4)_3/MoO_3$ catalyst and the two separate phases.
In this way, we can check if a kinetic isotope takes place on the
separate phases, and the measurements can be extended to a larger
temperature range more readily than under steady state conditions.
The pulses contained about 12% methanol in argon and 10% oxygen.

Results and Discussion

Pulses were taken at temperatures between 210 and 450 C at 30°
intervals, and in each case the conversion of methanol and the
product distribution were determined. Figure 5 shows a typical
result of an experiment. The intensity of various mass numbers is
plotted as a function of time. In this case 660 mg of the
commercial catalyst was in the reactor and twelve masses were
scanned at a rate of approximately 200 ms/scan, using the data
integrator. In most of the experiments, the scanning rate was
faster; 80-100 ms/scan for 8-10 masses.

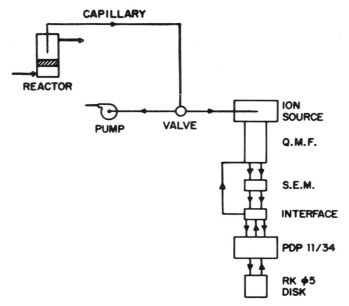

Figure 4.   On-line sample analysis.

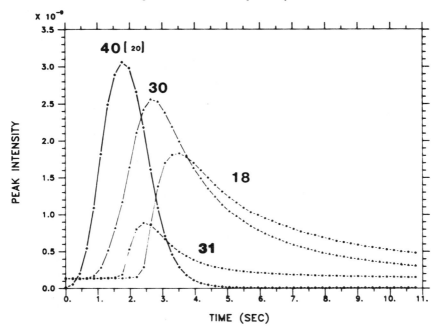

Figure 5.   Mass intensities vs. time using $Fe_2(MoO_4)_3/MoO_3$ at 270°.   Curve numbers indicate mass.   Key: 40, Ar/20; 30, CH_2O; 31, CH_3OH; and 18, H_2O.

The only products observed in the pulse experiments were
dimethylether, formaldehyde and water. This is a different
product distribution than in the continuous flow experiments and
indicates that dimethoxymethane and methylformate are not primary
products. In the analysis, peaks 18, 30, 31 and 45 are used for
water, formaldehyde, methanol and dimethylether, respectively.
Fragmentation patterns were also determined for the deuterated
species; they are nearly equal to that of CH_3OH. Small amounts of
carbon monoxide cannot be determined with great accuracy. The
figure shows clearly that all peaks have been delayed compared to
argon by several seconds. The peak for oxygen coincides with that
for argon, dimethylether coincides with methanol, and is somewhat
faster than formaldehyde. Water is adsorbed most strongly on the
catalyst. In blank experiments, the delay between peaks is not
more that 0.2 s. Figure 6 shows the delays of the tops of the
peaks for methanol, formaldehyde, and water relative to argon as a
function of temperature. As expected, there is a decrease with
increasing temperature, except for water which shows a minimum in
all cases.

Figures 7-9 show the fractional conversion of methanol in
the pulse as a function of temperature for the three catalysts and
the three methanol feeds. Evidently the kinetic isotope effect is
present on all three catalysts and over the complete temperature
range, indicating that the rate limiting step is the breaking of a
carbon-hydrogen bond under all conditions. From these
experiments, the effect cannot be determined quantitatively as in
the case of the continuous flow experiments, but to obtain the
same conversion of CD_3OD, the temperature needs to be 50-60°
higher. This corresponds to a factor of about three in reaction
rate. The difference in activity between MoO_3 and $Fe_2(MoO_4)_3$ is
larger in the pulse experiments compared to the steady state
results.

The conclusions on the rate limiting step are again supported
by the differences in product selectivity if completely deuterated
methanol is used; the selectivity to dimethylether relative to
formaldehyde is much larger. This is shown for the three
catalysts in Figures 10-12, in which the ratio of the amounts of
dimethylether and formaldehyde formed is plotted as a function of
temperature. In the case of CH_3OD, the water observed is a
mixture of H_2O, HDO, and D_2O, most of it being HDO.

A suggested reaction mechanism based on kinetic and isotope
experiments, product distributions, and the effect on these of
addition of water to the feed or leaving oxygen out is shown in
Figure 13.

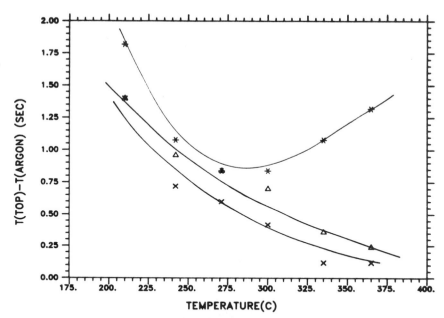

Figure 6.   Delay of top of peaks relative to Ar as a function of temperature. Key:
$\times$*, mass 34 (CD$_3$OD); △, mass 32 (CD$_2$O); and *, mass 20 (D$_2$O).*

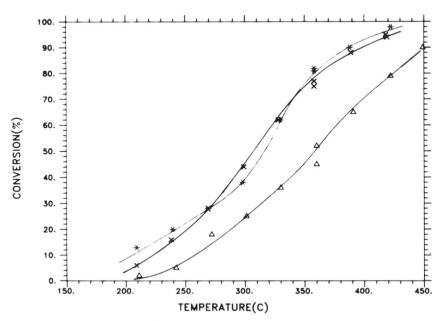

*Figure 7.   Conversion vs. temperature on 0.3 g of MoO$_3$. Key: △, CD$_3$OD; *,*
CH$_3$OD; and $\times$, CH$_3$OH.

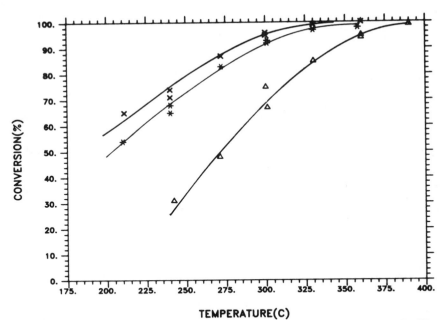

Figure 8.   Conversion vs. temperature on 0.3 g of $Fe_2(MoO_4)_3$.   Key: $\triangle$, CD_3OD;
$\times$, CH_3OD; and $*$, CH_3OH.

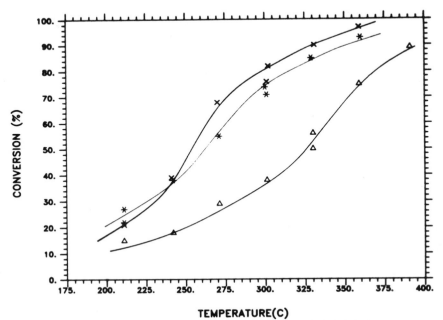

Figure 9.   Conversion vs. temperature on 0.2 g of $Fe_2(MoO_4)_3/MoO_3$.   Key: $\triangle$,
CD_3OD; $\times$, CH_3OD; and $*$, CH_3OH.

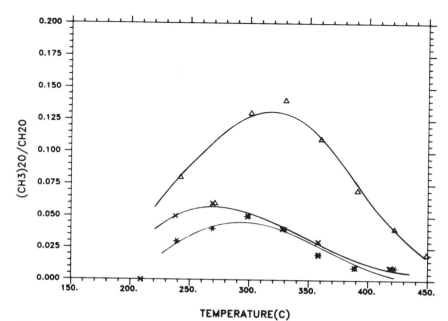

Figure 10.   Ratio of CH_3OCH_3 to CH_2O produced on 0.3 g of MoO_3. Key: $\triangle$, CD_3OD; *, CH_3OD; and $\times$, CH_3OH.

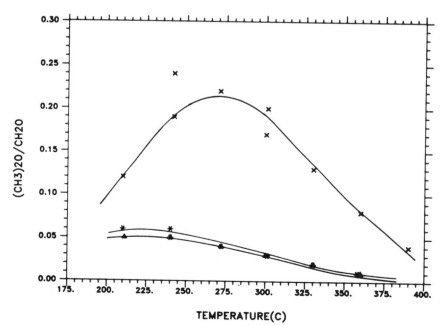

Figure 11.   Ratio of CH_3OCH_3 to CH_2O produced on 0.3 g of $Fe_2(MoO_4)_3$. Key: $\times$, CD_3OD; $\triangle$, CH_3OD; and *, CH_3OH.

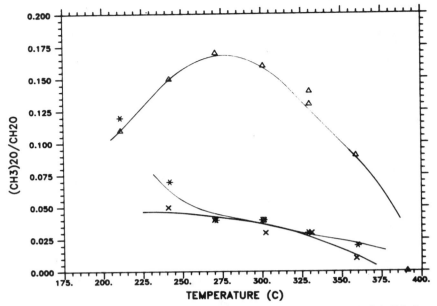

*Figure 12.   Ratio of CH_3OCH_3 to CH_2O produced on 0.2 g of $Fe_2(MoO_4)_3/MoO_3$.
Key: △, CD_3OD; ×, CH_3OD; and *, CH_3OH.*

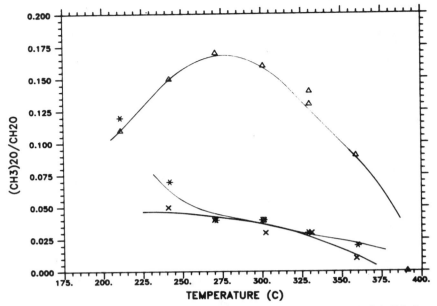

Figure 13.   Suggested reaction mechanism for CH_3OH oxidation on MoO_3.

A dissociative adsorption of methanol forming surface methoxy groups is suggested as the initial step. This is followed by the slow step, the formation of some form of adsorbed formaldehyde species. Evidence for the bridged species is not available, experiments with ^{18}O labeled methanol are expected to clarify this. Continued surface oxidation leads to a surface formate group and to carbon monoxide. All the byproducts can be obtained by combination of the appropriate surface species.

These initial experiments show that results can be obtained from this system that are comparable to those from the continuous flow reactor. The analytical system satisfies the requirements for accurate and rapid repetitive analysis. Scanning of 12 masses is possible at rates of approximately 100 ms/scan with good results. Further data manipulations are expected to yield additional results from this type of experiments.

Acknowledgement

The author is thankful to U. Chowdhry for preparing the pure phase ferric molybdate.

Literature Cited

1. Smith, J. M., Suzuki, M., Furusawa, T., Catal. Rev. 1976, 13(1), 43.
2. Bennett, C. O., Catal. Rev. 1976, 13(2), 121.
3. Kobayashi, H., Kobayashi, M., Catal. Rev. 1974, 10(2), 139.
4. Schweich, Villermaux, J., Anal. Chem. 1979, 51(1), 77.
5. Bernasek, S. L. Chadwick, P. A., Chem. Biomed. and Environ. Instrum., 1979, 9(3), 229.
6. Kasemo, B., Rev. Sci. Instrum. 1979, 50(12), 1602.
7. Winters, H. F., Raimondi, D. L., Grant, P. M., Clarke, D. C., IBM J. Res. Dev., July 1971, 307.
8. Latzel, J., React. Kinet. Catal. Lett. 1977, 7(4), 393.
9. Pernicone, N., Lazzerin, F., Liberti, G., Lanzavecchia, G., J. Catal. 1969, 14, 293.
10. Carbucicchio, M., Trifiro, F., J. Catal. 1976, 45, 77.
11. Van Truong, N., Tittarelli, P., Villa, P. L., Proceedings of the Climax Third International Conference on the Chemistry and Uses of Molybdenum, (H. F. Barry and P. C. H. Mitchell — eds.) Climax Molybdenum Company, Ann Arbor, Michigan, 1979, page 161.
12. "Catalyse de contact", ed. LePage, J. F., page 385, Editions Technip, 1978.
13. C. J. Machiels, submitted to J. Catal.

RECEIVED June 26, 1981.

Square Concentration Pulses in Oxidation Catalysis over Perovskite-type Oxides of Manganese and Iron

R. J. H. VOORHOEVE[1], L. E. TRIMBLE, and S. NAKAJIMA[2]
Bell Laboratories, Murray Hill, NJ 07974

E. BANKS
Polytechnic Institute of New York, Chemistry Department, Brooklyn, NY 11201

The objective of this work was to provide a technique for measuring catalytic reaction kinetics over oxides in a manner unaffected by hysteresis effects. Hysteresis is commonly introduced by changes in the stoichiometry of the catalyst in response to the reaction conditions (1). We wanted to measure the reaction kinetics in a time sufficiently short that the catalyst stoichiometry would not have changed between the beginning and the end of a series of measurements. To this end, it was necessary to substantially decrease the time on stream per data point, and the use of a pulse technique was therefore attractive.

In the commonly used pulse technique introduced by Kokes et al. (2), a spike of reagents is introduced into a carrier gas flowing over the catalyst and after that over a gas chromatography column which is used to determine the conversion of reagents to products. In this method, the concentrations of reagents in the catalyst bed are not defined, and kinetic measurements relating reaction rates to concentrations are impractical.

Our aim was to use a pulse sufficiently long so that a steady state concentration of reagents would be established (i.e., a flat-topped or "square" pulse), and to measure conversion by sampling a slice from that pulse into the gas chromatograph. Figure 1a shows the square concentration pulse desired, contrasted with the usual unshaped peak in Figure 1b. To enable us to "age" catalysts or to condition them, we also needed to provide a continuous mode corresponding to the traditional steady state rate measurement. Simple "flick of a switch" conversion between continuous and pulse mode operation would enable us to obtain a kinetic snapshot of a catalyst at any point during its life.

[1] Current address: Celanese Research Company, Summit, NJ 07901.
[2] Current address: Polytechnic Institute of New York, Chemistry Department, Brooklyn, NY 11201.

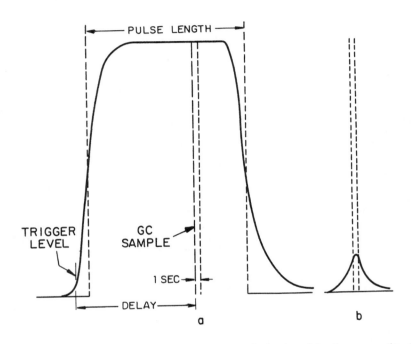

Figure 1.   *Pulse shapes in the square pulse method (a) and in the conventional pulse method (b). Solid lines give the pulse shape after passage through the catalyst. Trigger level refers to the point at which the pulse detector is triggered by the leading edge of the pulse.*

The incentive to modify our existing continuous-flow micro-unit to incorporate the square pulse capability was provided by our work on perovskite-type oxides as oxidation-reduction catalysts. In earlier work, it had been inferred that oxygen vacancies in the perovskite structure played an important role in catalytic activity (3). Pursuing this idea with perovskites of the type $La_{1-x}Sr_xFe_{0.5}Mn_{0.5}O_3$, our experiments were hampered by hysteresis effects which we assumed to be due to the response of the catalyst's oxygen stoichiometry to the reaction conditions. An example of such hysteresis is given in Figure 2 for the oxidation of CO over Sr_2MnFeO_6 and La_2MnFeO_6. Elimination of this hysteresis would enable us to collect a set of data descriptive of a constant state of the catalyst, and hence analyzable in terms of a true reaction order and activation energy.

In this work we attempt to measure kinetics data in a time short compared with the response time of the catalyst stoichiometry. An alternative is to measure kinetics in a true steady state, i.e., to increase the line-out time at each reactor condition until hysteresis is eliminated. The resulting apparent reaction orders and activation energies would be appropriate for an industrial mathematical model of reactor behavior.

Catalysts

The catalysts used in this study were La_2FeMnO_6 and Sr_2FeMnO_6 prepared by repeatedly mixing, ball milling, pressing and firing ceramic solids prepared from the corresponding carbonate raw materials (4). The pellets were fired in air at $1200^{\circ}C$ in a Pt boat in a tube furnace. The surface areas of the samples were measured by krypton adsorption in a Micromeritics automatic BET apparatus and were 0.29 m^2/g for Sr_2MnFeO_6 and 0.31 m^2/g for La_2MnFeO_6. Analysis of the samples by X-ray fluorescence showed a small contamination with Pt, presumably due to the firing in a Pt boat. The Pt level was not quantified.

Micro Flow Unit and Square Pulse Unit

The experimental set up (Figure 3) provides for continuous as well as pulse flow operation. Conversion between the two modes is instantaneous, limited only by the ~ 2 minute flush out time of the reactor. In the continuous flow mode, a mixture of CO, O_2 and He from a flow and ratio controller are fed over the catalyst via valve V_1 (AD) and then through GC sample valve V_4, by way of valves V_2 (EG) and V_3 (LN). The GC sampling is timed at regular intervals. After the sample has been taken, the reactor conditions are changed according to a pre-set pattern and the reactor is allowed to line out before the next sample is taken. In the pulse mode, a constant flow of He flows through valves V_5, V_1 (BD) over the reactor, through a pulse detector PD via valve V_2 (EH) and through the GC sample valve via valve

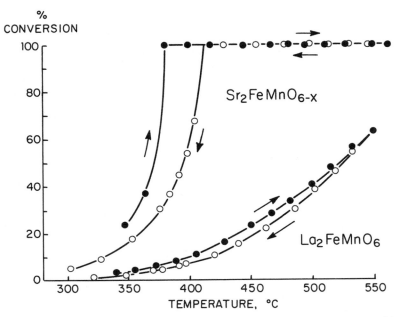

Figure 2. Continuous flow measurements of the conversion of CO over perovskite-type catalysts. The temperature is held constant for ~10 min for each data point. Conditions: 5% CO and 5% O_2 in He at a total flow of 100 mL/60 s (NPT); and catalyst weight 177 mg. Catalysts used as received.

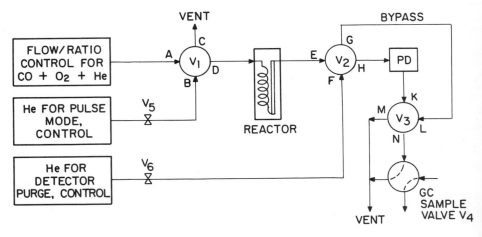

Figure 3. The square pulse method combined with a conventional continuous flow setup.

V_3 (KN). A pulse of the CO, O_2, He mixture is introduced by
switching valve V_1 from (AC, BD) to (AD, BC) for a predetermined
length of time, generally about 30 sec. After leaving the
reactor, the pulse is detected by PD at the trigger level (see
Figure la). The GC sample valve V_4 is at that time switched into
the "sample" position and after a preset delay a sample is injected into the GC carrier flow. The size of the slice taken
from the effluent pulse (1 sec in Figure la) is determined by the
volume of the sample loop in V_4. After the sample has been injected into the GC carrier flow, the conditions in the reactor,
e.g. the temperature, are automatically adjusted for the next
sample and data point. Figure 4 is a logic diagram of the pulse
mode operation. T_1-T_4 are event timers. V_1-V_4 are solenoid
valves. The reactor is heated by a fluidized sand bath with
internal exposed heating coil, allowing rapid changes of the
temperature.

Results

Hysteresis. Application of the pulse method to the measurement of the reaction rate of the oxidation of CO over La_2FeMnO_6
in pre-reduced and pre-oxidized states (top and bottom curves in
Figure 5, respectively) shows that hysteresis-free measurements
were made with 30 sec pulse length, 18-23 sec delay (see
Figure la) and 10-minute intervals between data points. Measurements by the continuous flow method in Figure 2 (bottom curve)
serve as comparison. The same duty cycle when applied to
measurements on Sr_2MnFeO_6 resulted in much reduced hysteresis
effects (Figure 6) as compared to the continuous flow measurements (Figure 2, top curves). The residual hysteresis was shown
to be due to bulk oxidation of the solid at temperatures of 550°C
and higher. It was observed that at 567°C, with a pulse flow of
5% CO and 5% O_2 in He completely converted to CO_2 and O_2, the
amount of O_2 absorbed by the catalyst in a single 30 sec pulse
was sufficient to increase the bulk oxygen content by 1%. Maintaining the Sr_2MnFeO_6 catalyst below 380°C eliminates hysteresis
completely, both in the pulse and in the continuous modes
(Figure 7). It was observed that the extent of hysteresis
depends also on the oxidation state of the catalyst, suggesting
that the rate of oxygen transport in the solid depends on its
stoichiometry. All data reported below were obtained after con-
ditioning the catalysts: a) by reduction in 5% CO in He at 600°C
for 1 hour, b) lining out in the pulse mode for 1 day at 340°C in
5% CO and 5% O_2 in He.

Elution of CO, O_2, CO_2 from the Catalyst. By varying the
delay time in Figure la, the elution of CO, O_2 and CO_2 can be
followed against argon as an internal non-adsorbing standard. CO
and O_2 are delayed by 1 second or less in a flow of 100 ml/min of
He over 177 mg of La_2FeMnO_6 at 490°C, while CO_2 is delayed by 2

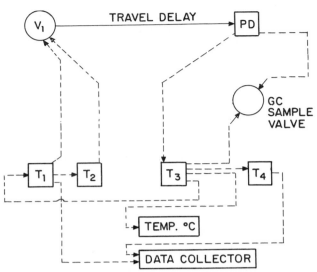

Figure 4.   Logic control diagram for the square pulse method as shown in Fig. 3.
− − → indicates causal directions.   Timer T3 sets the "delay" indicated in Fig. 1;
T4 prevents injection instabilities from registering in the data collection devices;
T1 determines the interval between pulses; and T2 determines pulse length.

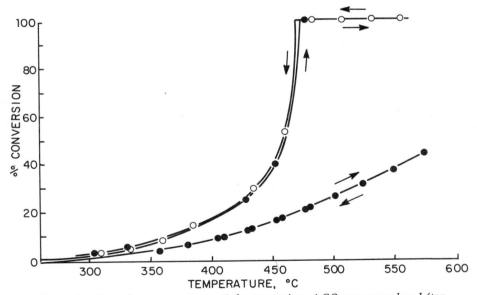

Figure 5.   Pulse flow measurement of the conversion of CO over prereduced (top
curves; reduction in 5% CO in He at 600°C for 1 h at a flow of 100 mL/60 s.)
and as received (bottom curves) La_2FeMnO_6. Duty cycle: 30 s pulse length, 10 min
interval, 18–23-s sample delay. Conditions: t, decreased or increased in steps; cata-
lyst weight, 176 mg; and flow, 100 mL/60 s of 5% CO and 5% O_2 in He.

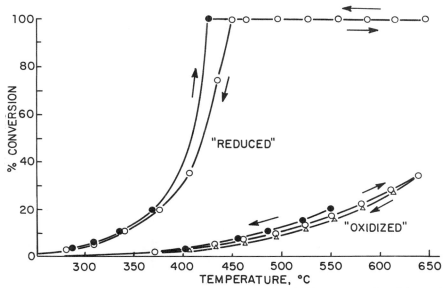

Figure 6. *Pulse flow measurement of the conversion of CO over prereduced (top curves; reduction in 5% CO in He at 600°C for 1 h at a flow of 100 mL/60 s) and as received (bottom curves) Sr$_2$FeMnO$_6$. Measurement as in Fig. 5. Catalyst weight, 177 mg.*

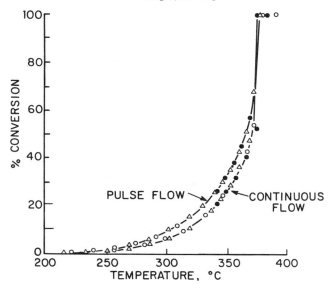

Figure 7. *Pulse flow and continuous flow measurements of the conversion of CO over prereduced (see Fig. 6 for conditions) Sr$_2$FeMnO$_6$. Flow of 100 mL/60 s of 5% CO and 5% O$_2$ in He over 177 mg of catalyst. Key: ●, points taken during stepwise decrease of temperature; and ○, △, during stepwise increase.*

seconds (Figure 8). Over Sr_2FeMnO_6 at 340°C, CO_2 is delayed 2
seconds under similar flow conditions. This illustrates that a
peak-shaped pulse as in Figure 1b would have suffered considerable chromatographic separation over the catalyst bed. It should
be noted that transient measurements of this type are much more
conveniently made by continuous sampling into a GC-MS
interface (5).

     Reaction Orders. Employing the same duty cycle of 30-second
pulses, delay of 13-23 seconds and 10-minute intervals between
pulses, we measured the effect of the partial pressures of O_2 and
CO on the reaction rates. The effect of the O_2 pressure is
given in Figure 9 for La_2FeMnO_6 and in Figure 10 for Sr_2FeMnO_6.
The effects of oxygen pressure are small: a five-fold increase
in pressure yields only a 30-50% increase in rate. The formal
reaction order at 320-325°C is +0.15 for La_2FeMnO_6 and +0.25 for
Sr_2FeMnO_6. The effect of the partial pressure of CO on the rate
over La_2FeMnO_6 (Figure 11) reflects a formal order of +0.60 at
308-320°C and an order of +0.4 at 340°C. The effect of the partial pressure of CO on the rate over Sr_2FeMnO_6 (Figure 12) shows
a more complicated behavior indicative of two different processes
yielding CO_2. At higher pressures of CO ($\geq$0.05 atm) a process
with nearly zero order in CO predominates. At partial pressures
of CO decreasing below 0.05 atm, rapidly increasing CO conversion
rates indicate the existence of a high-rate process which is
strongly inhibited by CO.

Discussion

     The objective of this work was to make hysteresis-free
measurements of oxide catalysts which are changing their oxygen
stoichiometry in response to the reaction conditions of the
measurement. This can be done if the time used for the measurement is much shorter than the relaxation time of the oxide
(stoichiometry is "frozen") or if the oxide stoichiometry equili-
brates sufficiently fast that true steady state measurements can
be made (6). In the intermediate case, non-equilibrium concentration profiles for oxide ion vacancies, oxide interstitials or
other defects are set up across the catalyst particles. These
concentration profiles will change with time in response to the
boundary conditions imposed by the reacting gas mixture. For the
duty cycle employed in the present work (30-second pulse, 10-
minute interval), the stoichiometry of La_2FeMnO_6 appears to be
"frozen" up to the highest temperature used, i.e., 575°C. In
contrast, Sr_2FeMnO_6 relaxes rapidly at that temperature, as shown
by changes of its oxygen content by 1% in a 30-second pulse. At
much lower temperatures, below 380°C, hysteresis-free measurements could be made on Sr_2FeMnO_6 in either the continuous or
pulse modes (Figure 7), suggesting a "frozen" stoichiometry.
However, the fact that the activity is different in the two modes

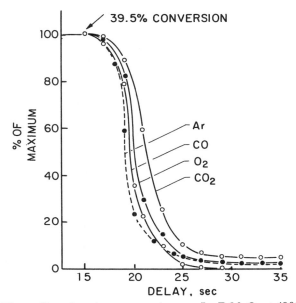

Figure 8.   *The trailing edge of a square pulse over La_2FeMnO_6 at 490°C. Ar was used as an internal standard. Feed gas was 5% CO and 5% O_2 in He at 100 mL/ 60 s. Data were normalized to 100% for the maximum level of the signal.*

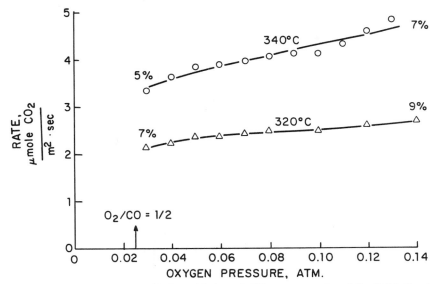

Figure 9.   *Reaction rate of the oxidation of CO over prereduced La_2FeMnO_6 at 320°C ($\triangle$) and 340°C ($\bigcirc$). See Fig. 5 caption for reduction. Dependence on partial pressure of O_2. CO partial pressure was constant at 0.05 atm. Percentages indicated at the extreme data points are CO conversions.*

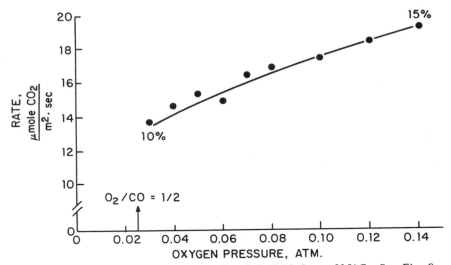

Figure 10. Reaction rate over prereduced Sr_2FeMnO_6 at 325°C. See Fig. 9 caption.

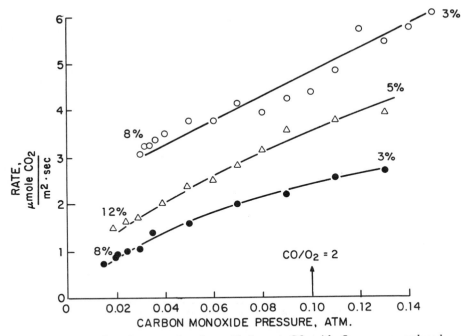

Figure 11. Reaction rate for the oxidation of CO with O_2 over prereduced La_2FeMnO_6 at 308° (●), 320° (△), and 340°C (○). See Fig. 5 caption. Dependence on partial pressure of CO. Oxygen pressure was constant at 0.05 atm. Percentages indicated with extreme data points are CO conversions.

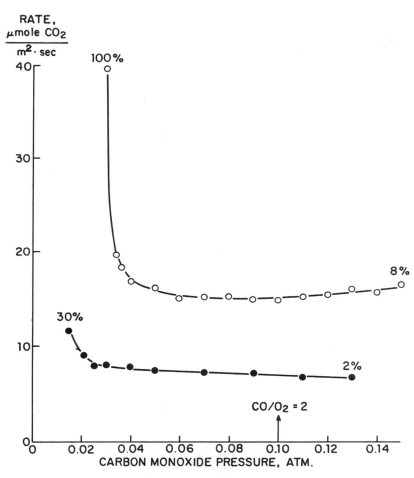

*Figure 12.   Reaction rate over prereduced Sr_2FeMnO_6 at 312°C (●) and 340°C
(○). See Fig. 11 caption.*

(i.e. about 30% higher in the pulse flow mode) leaves the possibility that relaxation of the near-surface region of the catalyst
affects the measurements of catalytic activity. Further work
will be necessary to elucidate this point.

The Sr_2FeMnO_6 and La_2FeMnO_6 perovskites show substantial
differences in the temperatures at which solid state relaxation
is observed. The data show that relaxation of Sr_2FeMnO_6 is
appreciably faster than that of La_2FeMnO_6. The catalytic activities are also quite different, with Sr_2FeMnO_6 more active by
about a factor 5. Dependence on the partial pressure of oxygen
is small, in accordance with the view that lattice oxygen is used
in the oxidation of CO (7, 8). The most striking difference was
found in the dependence of the reaction rate on the CO pressure
(cf. Figures 11 and 12). On Sr_2FeMnO_6, CO appears to be tightly
bound on the active sites (formal order $\leq$0) and we found some
preliminary evidence that CO might enter the solid, giving rise
to excess CO_2 being formed upon heating the catalyst to 570°C.
In contrast, La_2FeMnO_6 behaves more as expected for nearly-
stoichiometric oxides on which adsorption of CO is frequently
limited, with the formal reaction order (0.4-0.6) being in a
range frequently found for oxide catalysts (9).

Significant differences between Sr_2FeMnO_6 and La_2FeMnO_6 were
expected on the basis of their solid state properties. Stoichiometric Sr_2FeMnO_6 has formal valence states of Mn^{4+} and Fe^{4+}. The
latter is not stable, giving rise to complex defect structures in
the "parent" compound $SrFeO_{3-x}$ (10, 11). Sr_2FeMnO_{6-x} is
similarly expected to show defect structures including oxygen
vacancies at low x and tetrahedral rearrangements of oxygen
around Fe at higher values of x. Stoichiometric La_2FeMnO_6 has
the stable valence states Fe^{3+} and Mn^{3+}. High temperature (900-
1000°C) treatment in oxygen promotes the formation of Mn^{4+} and of
La-vacancies in the "parent" compound $LaMnO_{3+x}$ (12), with important effects on its catalytic activity (8). The present preparation (including firing at 1200°C in air) is expected to lead to a
more closely stoichiometric composition for La_2FeMnO_6. Solid
state studies of these compounds (4) will be related to their
catalytic properties in further studies now in progress.

The activities of the perovskite-type oxides are strongly
dependent on pretreatment in reducing or oxidizing atmospheres
at ~600°C. This was found for other perovskite catalysts as well
(1). Reducing pretreatments lead to more active catalysts
(Figures 5 and 6). The reason for this is not known, but better
binding of CO to the reduced surface is a possible explanation.

The striking rise of the reaction rate observed over
Sr_2FeMnO_6 as the partial pressure of CO decreases (Figure 12) was
attributed to an oxidation process inhibited by CO. This is an
intriguing result. Since this behavior might be due to contamination by low levels of Pt (from the Pt boat in which the ceramic
samples were fired), it will be necessary to study this feature
further with catalysts prepared without Pt hardware.

Conclusions

A square concentration pulse flow technique has been developed to study the kinetics of catalytic reactions over catalysts which change their stoichiometry in response to the reaction conditions. The technique makes it possible to obtain hysteresis-free kinetics data while greatly reducing the time during which the catalyst is exposed to the reaction mixture.

The technique has been applied to measurements of the kinetics of the oxidation of carbon monoxide over the perovskite-type catalysts Sr_2FeMnO_6 and La_2FeMnO_6. The pronounced defect nature of Sr_2FeMnO_6 was shown to give rise to higher activity, faster solid state relaxation and different kinetics than found for the more closely stoichiometric La_2FeMnO_6. Zero and negative reaction orders in CO indicate strong binding of CO to the defect catalyst.

Acknowledgements

We thank F. Schrey at Bell Laboratories for the surface area measurements and E. Bowers at Celanese Research Company for the Pt analysis. The work at Polytechnic Institute of New York was supported by the NSF (F. Stafford), Grant DMR-77-24133.

Literature Cited

1.   Voorhoeve, R. J. H.; Trimble, L. E.; Khattak, C. P.
     Materials Res. Bull. 1974, 9, 655.
2.   Tobin, H.; Kokes, R. J.; Emmett, P. H.   J. Am. Chem. Soc.
     1955, 77, 5860.
3.   Voorhoeve, R. J. H.; Remeika, J. P.; Trimble, L. E.; in
     Klimisch, R. L.; Larson, J. G., Editors; "The Catalytic
     Chemistry of Nitrogen Oxides"; Plenum: New York, 1955;
     p 215.
4.   Banks, E.; Berkooz, O.; Nakagawa, T.   NBS Special
     Publication 364, "Solid State Chemistry," Proceed. Fifth
     Materials Research Symposium, p. 265, issued July 1972,
     Washington, D.C.
5.   Happel, J.; Suzuki, I.; Kokayeff, P.; Fthenakis, V.
     J. Catal. 1980, 65, 59.
6.   The method of isotopic transient reaction kinetics (see 5),
     applied in a chemical steady state is an elegant application
     of the true steady state measurement.
7.   Keulks, G. W.   J. Catal. 1970, 19, 232.
8.   Voorhoeve, R. J. H.; Remeika, J. P.; Trimble, L. E.   Ann.
     N. Y. Acad. Sci.   1976, 272, 3.
9.   Yao, Y. F. Y.   J. Catal. 1975, 36, 266.
10. MacChesney, J. B.; Sherwood, R. C.; Potter, J. F.   J. Chem.
     Phys. 1965, 43, 1907.

11. Tofield, B. C.; Greaves, C.; Fender, B. E. F. <u>Mater. Res.</u>
    <u>Bull</u>. 1975, <u>10</u>, 737.
12. Voorhoeve, R. J. H.; <u>in</u> Burton, J. J.; Garten, R. L.,
    Editors; "Advanced Materials in Catalysis"; Acad. Press:
    N.Y., 1977, p 129.

RECEIVED July 28, 1981.

Reaction Rate Resonance in the Concentration Cycling of the Catalytic Oxidation of Carbon Monoxide

A. K. JAIN, P. L. SILVESTON, and R. R. HUDGINS

University of Waterloo, Department of Chemical Engineering, Waterloo, ON, Canada

Abstract

The effect of forced concentration cycling was investigated on the oxidation of CO over industrial V_2O_5 catalyst. The resulting rate, when time-averaged, exhibited frequency-dependent harmonic behavior, with multiple extrema. Some preliminary interpretation is provided by analogy to an electrical network containing resistance and inductance.

Some experimental studies (1-7) have demonstrated the possibility of improving the performance of a catalytic reactor through cyclic operation. Renken et al. (4) reported an improvement of 70% in conversion of ethylene to ethane under periodic operation. In a later article (2), they concluded that periodic operations can be used to eliminate an excessively high local temperature inside the catalytic reactor for a highly exothermic reaction. In our laboratory, Unni et al. (5) showed that under certain conditions of frequency and amplitude associated with the forced concentration cycling of reactants, the rate of oxidation of SO_2 over V_2O_5 catalyst can be increased by as much as 30%. Recently Cutlip (6) reported a thirty-fold increase in the rate of CO oxidation over Pt/alumina catalyst under forced concentration cycling at high frequencies. Abdul-Kareem et al. (7) observed large effects of the mean composition, period and the amplitude of cycling on the rate of oxidation of CO over V_2O_5. In this study, rate resonance was observed in the low frequency region (10 min < τ < 60 min). In the present work, the search for such a phenomenon was extended to higher frequencies (1 min < τ < 10 min), although the low frequency range was also re-examined. Symmetrical square concentration waves were used, as illustrated in Figure 1. The effects of amplitude and cycle period were investigated. Temperature (400°C) and space velocity (80,000 h^{-1}) were held constant. To identify processes which control behavior under cyclic operation, variations in rate during step-changes in gas phase concentration were studied at two temperatures: 390 and 440°C.

0097-6156/82/0178-0267$05.00/0

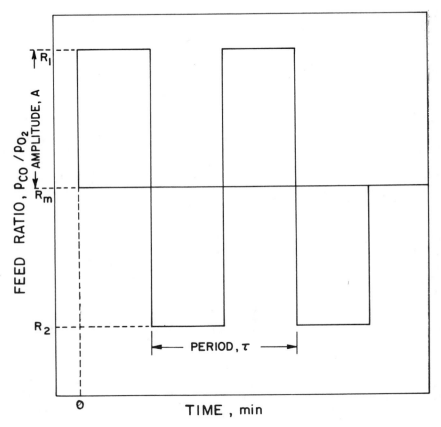

Figure 1.   Waveforms of concentration cycles used for forced periodic operation.

Experimental

The reaction was carried out in a differential flow reactor shown in Figure 2. It was packed with commercial vanadium catalyst supplied under the trade name "Aero SA" by Cyanamid of Canada Ltd. The catalyst contained 9.1% by weight vanadium reported as pentoxide, 10.1% potassium reported as oxide and 0.45% iron and was supported on diatomaceous silica. A specific surface are of 1.4 m^2/g and pore volume of 0.56 cm^3/g was measured in our laboratory (5,7). 30/40 U.S. mesh screen size particles were used to eliminate internal mass and heat transfer resistances. A total flow rate of 300 mL/min was used in all experiments to avoid external transport processes. The reactor was held isothermal by diluting the catalyst with an inert packing (glass beads) and immersing the reactor and preheating coil in a sand bath whose temperature was controlled to within 1°C by a proportional controller. All the cycling experiments were carried out at 126 kPa and 440°C.

Square-wave concentration cycling was achieved by means of an automatic timer, three-way solenoid valves and two flow circuits each containing fine-control needle-valves as shown in Figure 2. Thus for one half-period, high CO and low O_2 flow rates for a high CO/O_2 ratio were obtained; for the remaining half-period the flow was diverted by a solenoid valve and an automatic timer through a second flow circuit, to provide low CO and high O_2 flow rates for a low CO/O_2 ratio. The concentration of CO_2 in the product stream (containing O_2, CO, CO_2 and He) was recorded continuously by infra-red spectroscopy tuned to the peak of CO_2 at 2160 cm^{-1}. Mean values of rates shown in Figure 3 were found by integrating the spectrographic records.

Results and Discussion

The feed concentration of reactants was modulated about a mean CO/O_2 concentration ratio of 0.86 with amplitude A of 0.12 and 0.24. The instantaneous product concentration-time profile obtained during periodic operation is shown in Figure 4 for $\tau = 1$ min. During the first few cycles, the time-average rate in one cycle is increased followed by a decrease in the time-average rate until the cycle-invariant condition was achieved after approximately 70 cycles. When $\tau = 10$ min was employed, the maximum time-average rate was achieved in the first cycle and cycle-invariant condition was obtained after about 7-8 cycles. Thus the total time required to achieve the cycle-invariant condition was observed to be practically independent of the period of cycling. The time-average cycle-invariant rate was strongly affected by the period and amplitude as shown in Figure 3. The time-average rate was always higher than the corresponding steady state irrespective of the period of cycling. Resonance peaks appeared at $\tau = 2$ min, 20 min, and 40 min. The corresponding time-average rates were 150%, 100% and 60% higher than the steady-state rate under identi-

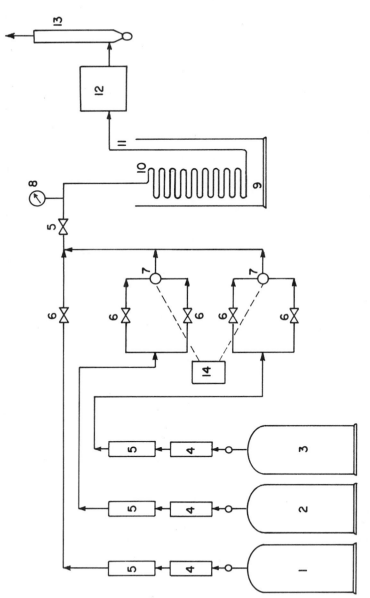

Figure 2. Schematic of the apparatus. Key: 1, He cylinder; 2, O cylinder; 3, CO cylinder; 4, molecular sieve dryer; 5, rotameter; 6, fine-control needle valve; 7, solenoid valve; 8, pressure gauge; 9, sand bath; 10, preheater; 11, reactor; 12, IR spectrometer; 13, gas bubbler; and 14, automatic timer.

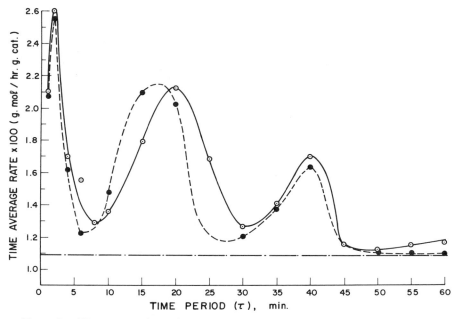

Figure 3.   Time-averaged reaction rate as a function of period. Key (concentration wave amplitude): ⊙, 0.24; ●, 0.12; and – – –, steady-state rate.

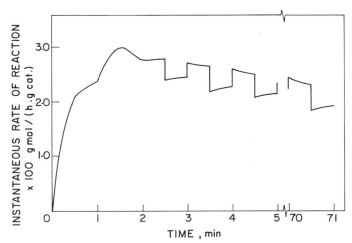

Figure 4.   Instantaneous CO₂ production as a function of time during periodic operation for τ = 1 min.

cal conditions of temperature and time-average concentration. The remarkable trend of the maxima with increasing τ resembles a damped harmonic sequence.

Above τ = 60 minutes, changes in the time-average cycling rate at an amplitude of 0.24 were too small to be significant. At the other end of the frequency spectrum, a different timer would have been required to achieve periods less than 1 minute. Periods of less than 1 minute were not investigated furthermore because at shorter periods the square wave becomes increasingly distorted by mixing in the system.

Figure 5 shows the response of the reaction system to the step-change in the gas phase composition from pure oxygen to CO/O_2 = 0.86. This experiment was conducted at two different bed temperature 440°C and 390°C. The figure shows that a transition period occurs between the start of the step-change and the final steady state. It is important to notice that the temperature has practically no effect on the length of the transition period, although the rate of reaction is changed eightfold. If surface relaxation were responsible for the transition period, about an eightfold change in the relaxation time over the fifty degree change in temperature used should have been seen. The change shown in the Figure 5 is very much smaller, implying that the process underlying the transition period has an activation energy of just a few kcal/mol. In contrast the activation energy for oxidation over this catalyst is about 39 kcal/mole. An exact value for the transition process cannot be calculated because the process has not been modelled.

A number of studies (9,10) have demonstrated that the valence of coordination of vanadium in typical vanadia catalysts changed with gas composition in contact with the catalyst. As we discuss below, changes at the surface are partially offset by transport of oxygen from interior strata of the catalyst. We believe it is the diffusion of oxygen, as an ion, which is responsible for the relaxation time observed.

The steady-state mechanism of CO oxidation over V_2O_5 catalyst is believed to involve adsorption of CO, possibly as carbonate ion on the oxidized catalyst site, decomposition of this ion releasing CO_2 and reduction of the co-ordination of vanadium at the site. The original co-ordination is restored by gas-phase oxidation. Apparently, lattice oxygen plays a direct role, while gas-phase oxygen only regenerates the reduced sites. The rate of CO_2 production is proportional to the concentration of CO and oxidized sites. Under cycling conditions, interior catalyst strata act as a reservoir for oxygen diffusing to or from the catalyst surface according to the steepness of the concentration gradient of lattice oxygen. This viewpoint is developed quantitatively in an as yet unpublished article (11). The resonance is attributed to constructive interference of the concentration cycling frequency and the natural frequencies of the system. Resonance in AC circuits requires resistance, conductance, and

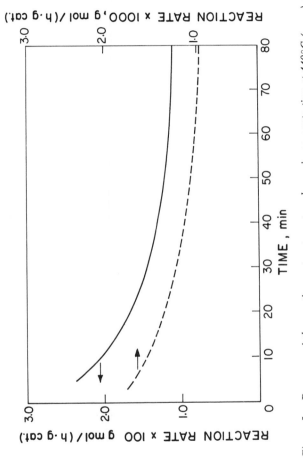

Figure 5. Response of the reaction system to a step change in concentration at 440°C (———) and 390°C (– – –).

inductance. Such circuits are an useful analogy to chemical reactions with forced cyclically varying reactant concentrations.
Resistance in the catalytic system occurs, we believe, in the
surface processes mentioned above, which are found in the reaction
sequence from adsorbed CO to adsorbed CO_2. Capacitance can be
assigned to the storage of unreacted adsorbed CO or some other
intermediate on the surface, whereas inductance results from the
transport of lattice oxygen between the interior of the catalyst
and the surface in response to the changing co-ordination of
vanadium at the surface. Models proposed by other investigators
of cyclic forcing of catalytic reactions (1,12-14) assume that ad-
sorption/desorption and surface reaction steps. These allow,
therefore, only for resistance and conductance. They do not predict resonance. Indeed, we show elsewhere (15-17) that they
even fail to predict rate improvement by cyclic operation.

Conclusions

Cycling reactant composition has been shown to significantly
increase the rate of oxidation of CO over a vanadia catalyst. The
time-average rate under cycling condition is strongly influenced
by the choice of amplitude and period of cycling. A plot of the
time-average rate vs. period exhibits multiple maxima and minima
suggesting a damped, harmonic response. The chemical relaxation
time of the reaction is shown to be practically independent of
the reaction rate and temperature. Diffusion of ionic oxygen
between the catalyst surface and the lattice is invoked in the
oxidation mechanism because it would provide inductance for the
catalytic reaction. Inductance is necessary to explain the resonance observations.

Legend of Symbols

A = amplitude of the change in concentration ratio, dimensionless
R_1 = ratio of partial pressure of CO to O_2 in CO rich half-cycle
       dimensionless
R_2 = ratio of partial pressure of CO to O_2 in CO poor half-cycle
       dimensionless
R_m = time-average ratio of partial pressure of CO to O_2, dimen-
       sionless
τ = period of the composition cycle, min.

Acknowledgements

Support for the work reported herein was provided through
grant A-2099 of the National Research Council (Canada). Catalyst
was provided by Cyanamid of Canada Ltd. Financial Support to
A.K. Jain through Dean of Engineering Scholarship by University
of Waterloo was appreciated.

Literature Cited

1. Denis, G.H. and Kabel, R.L., Chem. Eng. Sci. 1970, 25, 1057.

2. Renken, A., Müller, M., and Wandrey, C., Proc. 4th Int. Conf. Chem. Reac. Eng., 1976, p. 107.

3. Denis, G.H., and Kabel, R.L., AIChEJ, 1970, 16, 972.

4. Renken, A., Helmrich, H., and Schügerl, K., Chem. Ingr. Techn., 1974, 46, 647.

5. Unni, M.P., Hudgins, R.R., and Silveston, P.L., Can. J. Chem. Eng., 1973, 51, 623.

6. Cutlip, M.B., AIChEJ, 1979, 25(3), 502

7. Abdul-Kareem, H.K., Silveston, P.L., and Hudgins, R.R., Chem. Eng. Sci., 1980, 35, 2077.

8. Kadlec, B., Hudgins, R.R., and Silveston, P.L., Chem. Eng. Sci., 1973, 28, 935.

9. Mars, P. and Maessen J.G.H., Proc. Int. Gong. Catal. 3rd, 1965, 1, 264.

10. Boreskov, G.K., Polyakova, G.M., Ivanov, A.J., and Mastikhin, V.M., Dokl. Akad. Nauk SSR, 210 (3), 423 (Engl. transl).

11. Jain, A.K., Hudgins, R.R., and Silveston, P.L. Paper presented at ACS Symposium on Catalysis Under Transient Conditions. Las Vegas, Aug. 1980.

12. Mihail, R. and Paul, R., Chem. Eng. Sci., 34, 1058 (1979).

13. Horn, F.J.M. and Bailey, J.E., J. Optimiz. Theo. and Applic., 1968, 2, 441.

14. Bailey, J.E., "Periodic Phenomena", Chemical Reactor Theory: A Review (Edited by Lapidus, L. and Amundson, N.R.), Chap. 12, p. 785, Prentice-Hall, Englewood Cliffs, New Jersey, 1977.

15. Feimer, J.L., Jain, A.K., Hudgins, R.R. and Silveston, P.L., "A Critique of the Horn-Bailey Model of Forced Periodic Operation of Catalytic Reactors," paper submitted to Chem. Eng. Sci, 1980.

16. Jain, A.K., Hudgins, R.R. and Silveston, P.L., "Adsorption/ Desorption Models: How Useful to Predict Catalyst Behavior under Transient Conditions", paper submitted to Seventh North American Meeting, The Catalysis Society, Boston, 1981.

17. Jain, A.K., Hudgins, R.R. and Silveston, P.L., "Review of Bailey's Work on the Periodic Operations", paper submitted to Chem. Eng. Sci., 1980.

RECEIVED July 28, 1981.

Wavefront Analysis—A Special Transient Technique for Kinetic Investigations in Distributed Catalytic Systems

Application to the Water–Gas Shift Reaction

E. FIOLITAKIS and H. HOFMANN

Institut für Technische Chemie I der Universität Erlangen, Egerlandstr. 3, D-8520 Erlangen, Federal Republic of Germany

Abstract

The wavefront analysis of the reaction kinetics, a specific analysis of transient responses and a specific transient operation
of distributed heterogeneous systems, is here shortly presented.
The following results, obtained by the application of this technique for investigation of the low temperature watergas shift reaction, seem to be relevant for the transient operation under
technical operating conditions of catalytic systems consisted of
oxide-type catalysts. The existence of two microkinetic mechanisms of the formal Eley-Rideal type, i.e. having a linear dependence on p_{CO} and p_{H_2O}, has been found, besides a redox type
reaction between the ambient fluid and the catalyst which regu-
lates the oxygen activity on the catalyst surface. A systematic
investigation with the intension to find the dependence of the
microkinetics on definite oxidation states of the catalyst
(established by its pretreatment until equilibrium in a defined
H_2O, H_2, N_2 mixture) has revealed that for intermediary oxidation states and lower temperatures the redox reaction is slow
and the overall shift activity has a logarithmic dependence on
the pretreatment ratio p_{H_2O}/p_{H_2}. For extreme oxidation states and
higher temperatures, the redox reaction becomes faster and most
important than one of the microkinetic reaction steps of the formal E.R. type.

Heterogeneous catalysis under transient conditions encoun-
ters relaxation processes of the chemical conversion itself, pro-
ceeding by means of some surface intermediates, and relaxation
processes of the catalytic surface and/or the catalyst bulk caused by the changes in the composition of the ambient fluid (1, 19)
We like to name the first ones, following Temkin (1), the intrinsic or microrelaxations, having usually small relaxation times,
the last ones, the extrinsic or macrorelaxations, corresponding
in general to large relaxation times. In the case of macrorelaxations, we include here only the changes in the oxidation state of
catalyst, i.e. redox reactions between the ambient fluid and the

0097-6156/82/0178-0277$06.25/0

catalyst. For the oxygen activity distribution on different catalyst surfaces, a logarithmic dependence on the ambient fluid concentration has been found (2, 3, 5). This suggests in a first
approximation for the rate law of the macrorelaxations of this
kind an Elovich type dependence and in general a low activation
energy for low or high oxygen coverage on the catalyst surface
and a higher one for intermediary oxidation states. Therefore it
can be expected that at extreme oxidation states of the catalyst (highly oxidated or highly reduced) the macrorelaxations
proceed equally fast as the microrelaxations, whereas at intermediary oxidation states or/and near equilibrium the relaxation
times of the macrorelaxation would be large, maybe in the order
of magnitude of one hour.

Kinetic investigations of catalytic processes under transient conditions have to take into account this problem (see e.g.
(4), where the macrorelaxation of the redox type reaction has
been suppressed by means of a specific periodic operation). Kinetic expressions obtained by dynamic methods in general would give
a better understanding of the rate law than those obtained from
steady state measurements.

Apart from the above mentioned redox type reactions, we like
to consider (in connection with work to be published by us elsewhere) another type of relaxations, due to the possible reorgani-
sations of sorption intermediates on the catalyst surface, as
suggested by some investigations in our laboratory. This struc-
turing on the catalyst surface is equivalent to a change in the
entropy of the system catalyst surface / adsorbed intermediates
and seems to be responsible e.g. for the selectivity change under
transient conditions in the oxidation of hydrocarbons. Actually
this structural organization of the surface intermediates is also
a rate process which can be observed under transient conditions.

Our main motivation to develop the specific transient technique of wavefront analysis, presented in detail in (21, 22, 5),
was to make feasible the direct separation and direct measurements of individual relaxation steps. As we will show this objective is feasible, because the elements of this technique correspond to integral (therefore amplified) effects of the initial
rate, the initial acceleration and the differential accumulative
effect. Unfortunately the implication of the space coordinate
makes the general mathematical analysis of the transient responses cumbersome, particularly if one has to take into account the
axial dispersion effects. But we will show that the mathematical
analysis of the fastest wavefront which only will be considered
here, is straight forward, because it is limited to ordinary differential equations; dispersion effects are important only for
large residence times of wavefronts in the system, i.e. for slow
waves. We naturally recognize that this technique requires an
additional experimental and theoretical effort, but we believe
that it is an effective technique for the study of catalysis
under technical operating conditions, where the micro- as well as
the macrorelaxations above mentioned are equally important.

The elements of the wavefront analysis

The transient technique of wavefront analysis is based on one hand on a specific analysis of the transient responses, on the other hand also on a specific transient operation and is applicable only to distributed heterogeneous systems, e.g. a heterogeneous catalytic packed bed reactor (see Figure 1). By wavefront is meant here a disturbance, e.g. a step disturbance in the composition at the inlet of the system which preserves more or less its identity and propagates at a definite speed to the exit of the system. The only mechanism considered here, responsible for the origin and propagation of the wavefront is that of the convective accumulation. By this, we mean that there are at least two phases, one movable and the other one fixed, and we have to deal with the transfer of energy (heat) and material between the phases and simultaneously with the chemical conversion. The wave propagation considered here, is thus connected with the energy (heat) and mass capacitances of the phases; we like therefore to name this "capacitive waves". The propagation speed corresponds to the ratio of the respective total capacity (of heat or mass) to the respective capacity of the movable phase. Dissipation effects like dispersion, heat conduction etc. are not considered here, because they have been suppressed conveniently by means of a specific operation (e.g. choosing relatively large flow rates) in order to have a small influence on the transients around the wavefront.

The specific interpretation of the transient response consists in separating it into two parts.
- the wavefront strictly speaking. For an observer on the wavefront, the change of the amplitude of the disturbance is noticed as the relaxation along the propagation line and is named here the space relaxation.
- The remaining transient response after passage of the wavefront, it is the transient beyond the wavefront, as observed by an observer fixed at a definite position in space, named here time relaxation.

The elements of the wavefront analysis are:
· the propagation speed of the wavefront
· the space relaxation of the wavefront
· the time gradient of the time relaxation at the wavefront.

Basic equations

The following Eqns. are important for the mathematical analysis of the elements of the transient responses around the wavefront (see ($\underline{5}$, $\underline{6}$). In these equations y_i stays for the state variables, such as concentration in the ambient fluid, or on the catalyst surface, temperature etc., f_i is the relaxation function (e.g.

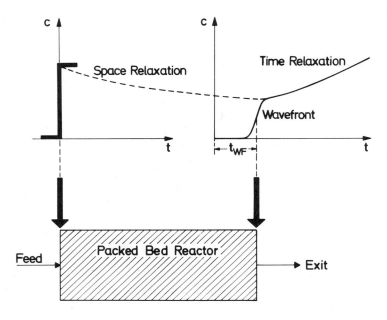

Figure 1.   One-dimensional representation of the propagation of a concentration step-disturbance in a packed bed reactor (5).

chemical rate) and f_{ij} is the partial derivative $\partial f_i/\partial y_i$. We assume that a step inlet disturbance in the variable y_1 has been propagated in the system, the propagation line of the wavefront $\alpha_o(t,z) = 0$ separates the undisturbed from the disturbed region in the propagation field. The varible y_1 changes discontinuously on α_o (space relaxation) and other variables are constant on α_o.

Space relaxation

$$(y_1)_{\alpha_o} = \int_0^{\tau_o} (f_1)_{\alpha_o} \, dt'$$ (1)

Time relaxation

$$\left(\frac{\partial y_i}{\partial t}\right)_{\alpha_{o+}} = (f_i)_{\alpha_o} \quad , \; i = 2, \ldots, N$$ (2)

$$\left(\frac{\partial y_1}{\partial t}\right)_{\alpha_{o+}} = \left[f_1(\tau_o)\right]_{\alpha_o} \int_0^{\tau_o} \frac{\sum\limits_{i=2}^{N}\left[f_{1i}f_i\right]_{\alpha_o}}{\left[f_1(t')\right]_{\alpha_o}} \, dt'$$ (3)

Propagation speed

$$\left(\frac{dz}{dt}\right)_{\alpha_o} = \frac{1}{\lambda_1} \frac{\partial (uy_1)}{\partial y_1}$$ (4a)

for a simple wavefront

$$\left(\frac{dz}{dt}\right)_{\alpha_o} = \frac{\left[uy_1\right]}{\left[\lambda_1 y_1\right]}$$ for a shock wavefront (10) (4b)

where u is the average velocity in the convective direction and λ_1 the capacity parameter, the brackets meaning discontinuities.
   For the special case of linear relaxation functions f_i or when the integrand in Eq. (3) is invariant on the wavefront, we have the simplified Equation for the time relaxation of the propagating disturbance:

$$\left(\frac{\partial y_1}{\partial t}\right)_{\alpha_{o+}} = \tau_o \sum_{i=2}^{N} \left[f_{1i}f_i\right]_{\alpha_o}$$ (5)

Application principles of wavefront analysis

As for any transient technique, it is essential also for wavefront analysis

a) to ascertain (by preliminary experiments) mechanistic information by means of a qualitative and semi-quantitative analysis

b) to perform systematic experiments in order to obtain accurate numerical values for the kinetic parameters by means of a quantitative analysis.

The qualitative analysis consists e.g. in stimulating pure sorption steps by using a step inlet disturbance in the fluid concentration of the corresponding species. The responses are different depending on whether the step is kinetically controlled or not (see Figure 2, 3).

In the first case, the following equation, derived from (5), is important for a good adjustment of the individual relaxations (see (5)).

$$\frac{\tau_{c,time\ relx}}{\tau_{\theta,time\ relx}} = \frac{\tau_{c,space\ relx}}{\tau_o} \frac{\theta_{eq}}{\theta_{max}} \tag{6}$$

which links the time constant of time and space relaxations with the maximal and equilibrium internal storage.

If the step is only equilibrium controlled, a lumped analysis of both fluid and solid phase is possible and the corresponding nonlinear wavefront analysis is straight forward too. Analysis of propagation speed data yield information about the relevant equilibrium sorption mechanism (7). The different form of the transients when a reaction step has been stimulated, is discussed in the Appendix.

Application of wavefront analysis to watergas shift reaction

We report here about the investigation of the low temperature watergas shift reaction on an industrial catalyst (GIRDLER G 66-B and E with copper and zinc oxides as main components) under transient conditions by means of wavefront analysis. After a qualitative analysis to obtain information about the relevant mechanistic scheme the main effort has been concentrated on the dependence of the microkinetics on different oxidation states of the catalyst. The watergas shift reaction in its overall formulation

$$CO + H_2O \rightleftharpoons H_2 + CO_2 \qquad -\Delta H = 42,7\ kJ/mole$$

is a simple reaction without any selectivity problems. The investigations have been mainly performed under atmospheric pressure and in the temperature interval 180 to 280°C. The order of magnitude of the overall relaxation time of the microkinetics is about 1 second. For the macrorelaxations of the redox type, also included here (9), it is important that for extreme oxidation states of

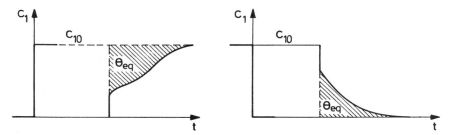

Figure 2. Transient responses when stimulating pure sorption step that is kineti-cally controlled (5). Key: left, loading; and right, unloading the storage.

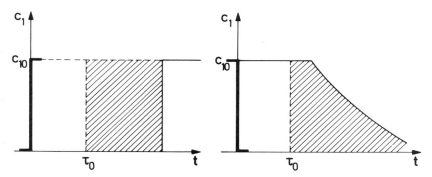

Figure 3. Transient responses when stimulating a pure sorption step that is not kinetically controlled (5).

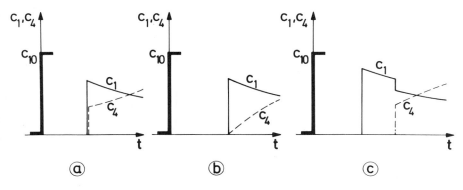

Figure 4. Transient responses when the reaction step $A_1 + \ldots \to A_4 + \ldots$ has been stimulated (see Appendix) (5).

the catalyst, the conversion at the wavefront (that means the space relaxation) contains significantly both micro- and macrorelaxations. The time relaxation reflects the change of the surface potential with time (i.e. the change of the oxidic active sites with time).

The experimental equipment used in these investigations and the analytic system, are described in detail elsewhere ($\underline{5}$, $\underline{7}$).

Qualitative and semiquantitative analysis
Typical transient responses. The transient responses of the four reactants after a simultaneous step change in the inlet concentration of both CO and H_2O for a catalyst being initially in N_2 at 130°C are depicted in Figure 5. The CO response is the fastest one and behaves practically like a step response (the ascending signal has a time constant of about 5/4 seconds which is nearly equal to the time constant of the gas inlet system to achieve a continuous sampling from the normal pressure reactor vessel to the high vacuum chamber for mass spectrometric analysis; the lower operating temperature has been selected here in order to obtain stronger chromatographic separation effects). The same is also valid for the ascending CO_2-signal, but the last one is shifted parallel to the CO-front by a time interval of about 10 seconds, obviously caused by the retention of CO_2 on the solid capacitance. The CO_2 transient exhibits a maximum on the wavefront, whereas the CO transient exhibits a minimum on the wavefront. The H_2O wavefront has a larger retention time of about 30 s compared to the CO wavefront. As independent investigations with faster H_2O-fronts confirmed the H_2O transient as a shockfront, the H_2O transient here observed would represent a "blurred shockfront", because of the longer influence of dispersion effects (10). The maximum in the CO_2 concentration on the wavefront has been obviously caused by the partial reduction of the catalyst surface by CO, that means consumption of a part of the oxygen capacitance of the catalyst, whereas a possible subsequent reoxidation of the catalyst by H_2O seems to proceed slowly. The H_2 response in this experiment is certainly not a step response, it starts ascending from zero at the same time as the CO front appears, suggesting that the CO front and the H_2 front propagate with the same velocity. As during the pretreatment of the catalyst no H_2 and H_2O was present, the H_2 production is generated only by the shift reaction. The fact that CO_2 is not measured at the reactor exit for a certain period of time, for which H_2 has been observed, is caused by the large CO_2 capacitances of the catalyst, giving rise to the CO_2 shockfront. The slowest shockfront, it is this of H_2O represents in this case the moving boundary, behind of which the shift reaction only can take place. Actually the slow H_2O wavefront implies that the H_2 signal is ascending slowly, but in any case the reduction of the catalyst by CO (suggested also by the small stoichiometric deviation of CO_2, H_2 in the intermediary steady state) proceeds immediately behind the CO wavefront, independent

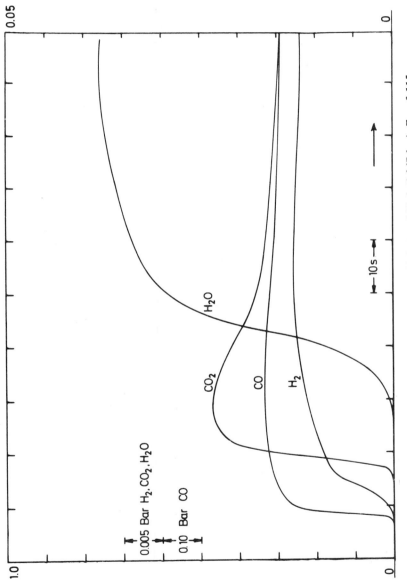

Figure 5.   Transient responses after a step in CO (0.333 bar) and H_2O (0.067 bar), $Z = 0.115$ m, 60 g at 130°C. Catalysator Girdler G-66B, 0.11 Nm^3/h (5).

of the presence of H_2O. Independent investigations have also confirmed that H_2O inhibits the reduction of the catalyst by CO (see
(7)). The case handled here, is a typical example, where more
quantitative information can be gained only by the nonlinear wavefront analysis. But in the investigations which are reported below the experimental conditions have been always selected so that
the effects of the nonlinear wave propagation discussed above
(e.g. the slowly ascending H_2-signal) do not complicate the kinetic investigations.

Furthermore for the relevant adsorption steps: CO-adsorption,
H_2O-adsorption, CO_2-desorption, H_2-desorption, one concludes from
the form of the transients that the first three are not kinetically controlled, as they are shock fronts and that H_2 has no measurable sorption capacitance in presence of CO. The last conclusion
has been also confirmed by independent experiments, i.e. strip-
ping-off of adsorbed H_2 by CO, and the transients in this case
suggest that also the H_2 desorption, at least under the selected
experimental conditions, is not kinetically controlled. Independent transient experiments showing that the CO_2 sorption transients can have the form of shock wavefronts or simple wavefronts
(see (7)) support the conclusion that CO_2 sorption is not kinetically controlled; it could be also shown that H_2O can strip off
CO_2.

Sorption of H_2O. A systematic investigation of the H_2O sorption by analysing the wave propagation data has implied that H_2O
sorption proceeds by two mechanisms, one of which follows a Langmuir isotherm ($-\Delta H = 33,5$ kJ/mole) and the other one, probably
chemisorption or capillary condensation in the micropores of the
catalyst, has a capacitance which is saturated even at low partial
pressures of H_2O (see (7)).

Sorption intermediates of CO and H_2O in the shift reaction.
Additional information regarding the participation of the sorption intermediates of CO and H_2O in the shift reaction have been
established by the following experiments:
The catalyst has been pretreated for several hours with a CO/N_2
mixture to attain a uniform and, as shown, low shift activity.
Thereafter a periodic operation has been established as follows:
testing time 1 min with a H_2O/N_2 mixture, recuperation time of 30
min with the same CO/N_2 mixture as before.

The transients shown in Figure 6 (see (9)) suggest that in
the H_2O/N_2 phase, H_2O reacts with adsorbed CO to produce CO_2 and
H_2 and that the H_2 wavefront concentration is two-fold of the CO_2
wavefront concentration. The overproduction of H_2 on the wavefront
seems to have been caused by the reoxidation of the catalyst by
H_2O. The reoxidation rate is therefore on the wavefront equal to
the rate of the shift reaction, it is namely the limiting step in
the global relaxation process. Furthermore, the fact that $H_2:CO_2$
is 2:1 on the wavefront suggests that presumably the shift con-

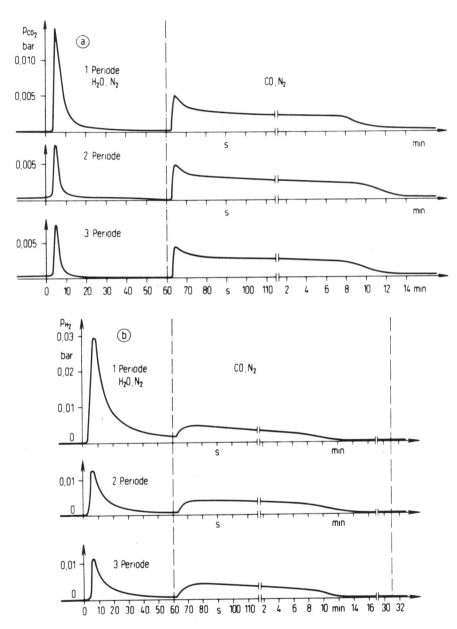

Chemie Ingenieur Technk

Figure 6.   CO₂, H₂ transients for a periodic operation at 220°C for a catalyst initially intensively reduced about 18 h in CO/N₂ (9).

version is not possible for a catalyst surface at low oxidation
state (intensively reduced catalyst). It is only possible after
some structures of oxygen on the surface have been built up, e.g.
some oxide forms of copper. In the subsequent 2nd period, the
shift activity on the wavefront is lowered by the factor of about
2, that means a time constant for CO-recuperation of the initial
activity of about 43 min (time constant = 30 min/ln2).

During the CO/N_2 phase, one obtains a long ranging low production of both CO_2 and H_2 in stoichiometric ration and at practically constant rate. This kind of responses suggest that the
stimulated H_2O sorption capacitance is much larger than the CO capacitance and that this part of H_2O sorption must take place on
surface sites different from the shift reaction sites, from where
H_2O is probably transported by a slow mechanism to the shift reaction sites. The limiting step during this period is obviously
the transfer of H_2O.

In order to clear up whether the sorption intermediates of
CO and H_2O participating in the shift reaction are reversibly or
irreversibly adsorbed a similar experiment has been accomplished
with an additional stripping period of 30 min with purified N_2,
enclosed between the testing periods. As it could be shown, this
stripping phase was sufficient to remove the adsorbed H_2O but not
the CO (see (9)), therefore we like to regard the H_2O sorption as
reversible and the CO sorption as irreversible.

Transient behavior of the catalyst at high oxidation state.
The significance of the redox type reactions between reaction medium and catalyst is depicted in Figure 7, where the conversion
of CO to CO_2 has been stimulated under shift conditions. After a
long range pretreatment of the catalyst in a mixture of H_2O/N_2
(0,30 bar H_2O) a periodic operation has been established consisting of a testing phase of 1 min with a mixture of CO, H_2O, N_2
(0,30 bar CO, 0,30 bar H_2O) at $160°C$ followed by a recuperation
phase with the same H_2O/N_2 mixture of 12 min. The transients of
both H_2 and CO_2 show in the concentration phase plan a hysteresis
loop, caused obviously by the redox reaction between the reaction
medium, mainly CO, and the catalyst. In the consecutive periods,
the hysteresis loop becomes smaller and smaller and vanishes practically after the fourth period. Furthermore, the intermediate
steady state of the shift conversion decreases from the first to
the fifth period, suggesting its dependence on the oxidation state
of the catalyst, as it changes from period to period. Also in this
case, one can expect that the recuperation time constant is in the
order of magnitude of several minutes, as the 12 min exposure of
the catalyst to H_2O is not sufficient to retain the initial redox
and shift activity.

Transient behavior of the catalyst at low oxidation state.
After a reduction of the catalyst in a CO/N_2 mixture, the transient shift operation, depicted in Figure 8 (see also (7)), re-

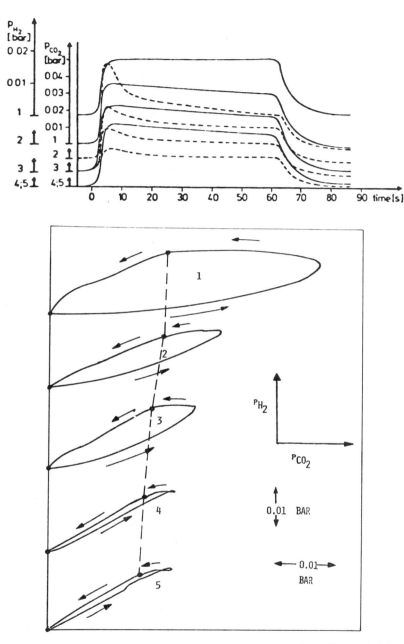

Figure 7. CO₂, H₂ transients (top) and reaction trajectories (bottom) for a periodic operation with the catalyst initially at high oxidation level (7). Key: ——, P_{H_2}; and – – –, P_{CO_2}.

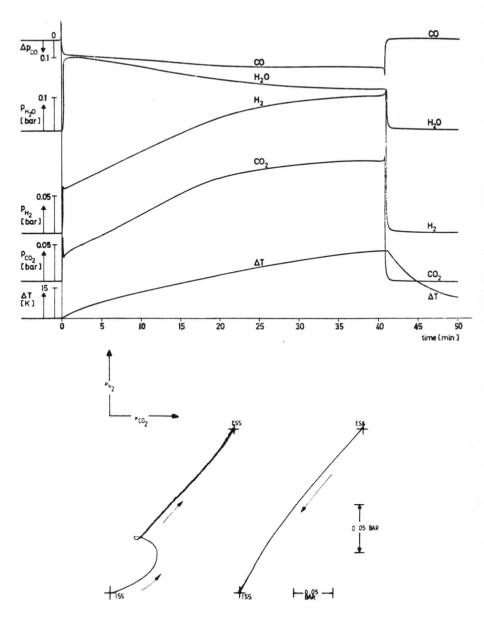

Chemical Engineering Science

Figure 8.   Transient responses (top) and reaction trajectories (bottom) with the catalyst initially at low oxidation level (7). Key: ISS, initial steady state; and ESS, end.

sulted in an increase of the overall activity. This has been established by a subsequent testing with the same CO, H_2O, N_2 mixture after cooling of the reactor in a N_2 stream, where the wavefront concentration of CO_2, H_2 was higher than the corresponding one in Figure 8. This different behavior (compared to Figure 6) has presumably its cause to the fact that there the catalyst, because of the intensive reduction, has been presumably deactivated through brass formation (see (11)). But also in this case, the wavefront concentrations of H_2 and CO_2 behave like 2:1 and the reaction trajectory in the phase plane approaches only asymptotically the stoichiometric line 1:1 (as the concentration transients become stationary after 30 min, the time constant of this relaxation process amounts $30/4 \approx 7,5$ min).

Semiquantitative analysis. In a preliminary investigation, the analysis using Eq. (1) of space relaxation data, obtained by testing under shift conditions the catalyst at high oxidation state by a CO, H_2O, N_2 mixture at constant $p_{H_2O} = 0,40$ bar and various p_{CO} and at different temperature levels between 180°C and 260°C, resulted in the following kinetic parameters for the power law rate $r = k \cdot p_{CO}^n$:

$$n = 0,625 \pm 0,007, \quad E = 46,270 \pm 0,147 \text{ kJ/mole} \quad (\text{see } (7))$$

The nonlinear dependence of the reaction rate on the partial pressure of CO suggests that there are stimulated either a single reaction step with the established nonlinearity or at least two steps, one of which has a linear and the other one a nonlinear ($n < 1$) dependence on p_{CO}. To clear up this question, a periodic operation of the reactor at the 180°C level and at a middle oxidation state (pretreatment at $p_{H_2O}/p_{H_2} = 5,0$) has been accomplished (see Figure 9) with a H_2O/N_2 testing mixture followed by a CO/N_2 recuperation mixture. After an intermediate activity of the catalyst has been attained by this periodic operation, CO has been added in the testing mixture.
The transients of Figure 9 suggest that the wavefront conversion, when both CO and H_2O are present in the gas phase, is approximately equal to the sum of the wavefront conversions in the CO/N_2 and H_2O/N_2 phase. (At the established pretreatment T = 180°C, $p_{H_2O}/p_{H_2} \cong 5,0$ are relevant sorption effects of CO_2 but not of H_2; thus only the H_2 wavefronts represent rather the shift conversion). Therefore it seems conceivable that there are two different mechanisms which participate in the CO shift conversion which is also in agreement with the established two different sorption mechanisms for H_2O and with the transient behavior, depicted on Figure 6.

Quantitative Analysis. A systematic kinetic investigation has been performed therafter as follows: The catalyst has been pretreated for several hours with a mixture of $H_2O/H_2/N_2$ at defi-

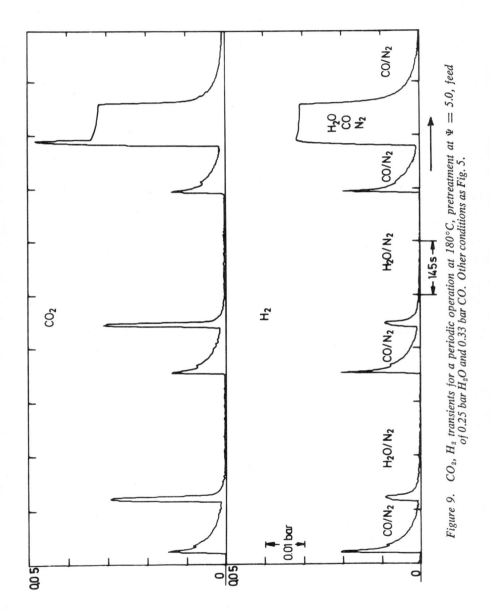

Figure 9. CO_2, H_2 transients for a periodic operation at 180°C, pretreatment at $\Psi = 5.0$, feed of 0.25 bar H_2O and 0.33 bar CO. Other conditions as Fig. 5.

nite levels of the ratio $\Psi = p_{H_2O}/p_{H_2}$, until equilibrium has been established. The Ψ has been variied over three decades, from 0,01 to 10,0. The shift activity of the catalyst has been tested thereafter by a mixture of CO, H_2O, N_2 with the inlet step disturbance of CO and H_2O being variied between 0,062 to 0,333 and 0,i34 to 0,667 bar at the temperature levels 230, 250, 280°C. A series of experiments at the 230°C level has been performed earlier and in the meantime the reactor was shut down for about three months and stood in contact with air at room-temperature. This "resting" period resulted obviously in a higher overall activity of the catalyst as established by the conversion on the wavefront (see (12)).

In both cases the fitting of the space relaxation data with a single Langmuir-Hinshelwood kinetics as well as with a single stationary redox kinetics (i.e. assuming that on the wavefront the microkinetics are in quasi steady state) has failed (see (3, 12)). On the other hand, the attempt to fit the data has given the hint that there must exist a practically linear dependence of the rate on both, p_{CO} and p_{H_2O}. If one takes into consideration the results of the qualitative analysis, in particular the fact that there are two microkinetic mechanisms, by which the shift reaction proceeds under the participation of sorption intermediates, it seems plausible that these mechanisms are of the formal type of the Eley-Rideal mechanism; whether the linear dependence on the p_{CO}, p_{H_2O} has its origin in the reaction of gas phase CO and H_2O, or of linearly adsorbed CO and H_2O with chemisorbed H_2O and CO, respectively, is still an open question (it can exist for example a "linear" weak adsorption accompanied by a second stronger adsorption after Wicke (18). Therefore the following mechanistic scheme seems to describe the low temperature watergas shift reaction under transient conditions (neglecting the reverse reaction, as here $K_p \approx 100$)

(i) $H_2O + (CO)^* \longrightarrow H_2O + CO_2$, where $(CO)^*$ is CO chemisorbed

(ii) $CO + (H_2O)^* \longrightarrow H_2 + CO_2$, where $(H_2O)^*$ is H_2O chemisorbed

(iii) Redox mechanism regulating the surface oxygen activity

(a) $CO + (O) \longrightarrow CO_2 + (\ )$, where (O) is reducible surface oxygen

(b) $H_2O + (\ ) \longrightarrow H_2 + (O)$

The fitting of space relaxation data using Eq. (1) to this mechanistic scheme (space relaxation data are always isothermal, because transient temperature effects are not relevant for the amplitude change of a concentration disturbance; this is just an advantage of wavefront analysis of reaction kinetics), reported in (3, 5, 12), supposing a Langmuir type chemisorption for (CO) and (H_2O) has confirmed that (see Figure 10, 11):

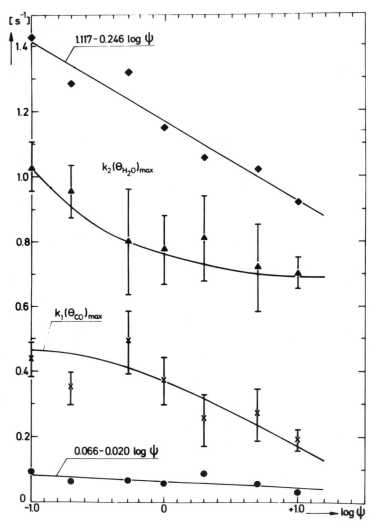

Figure 10. Dependence of the effective microkinetic parameters $k_i\,\theta_i$ and of the eigenvalues of the space relaxation on log Ψ for 230°C, first investigation (3). Key: $\times \triangleq k_1'$, Eley-Rideal, CO adsorbed; $\blacktriangle \triangleq k_2'$, Eley-Rideal, H_2O adsorbed; $\bullet \triangleq m_1$; and $\blacklozenge \triangleq m_2$. Conditions: $\Psi = P_{H_2O}/P_{H_2}$ in pretreatment phase at 230°C.

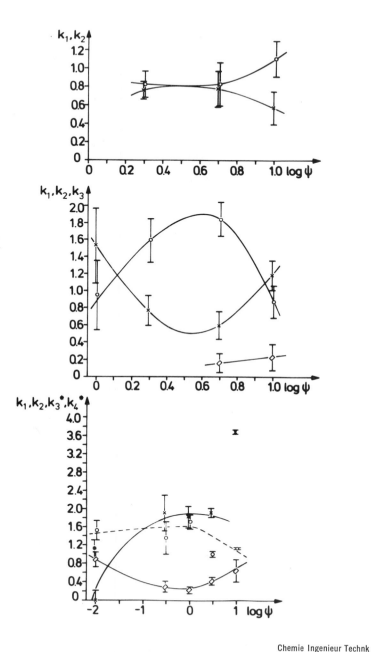

Chemie Ingenieur Technk

*Figure 11.   Dependence of the microkinetic parameters on log Ψ, second investigation (12).   Key: ○, k_1; ✕, k_2; □, k_3^*; *, k_4^*; top, 230°C; middle, 250°C; and bottom, 280°C.*

a) there is a significant dependence of the kinetic parameters on
   the oxidation level of the catalyst
b) for intermediary oxidation states the redox parameters are not
   significant and highly correlated; for low oxidation states, a
   shift in the mechanistic scheme occurs that means one of the
   Eley-Rideal mechanisms vanishes completely in favour of the
   redox mechanism (Figure 11, 280°C)
c) the linear dependence of the largest eigenvalue (corresponding
   here to the overall shift activity (from the space relaxation
   data) on log γ depicted in Figure 10 (see (3, 5)) suggests
   that for the oxygen distribution on the catalyst surface (oxygen activity) a logarithmic dependence on the gas phase partial pressure exists, i.e. a chemisorption isotherm of Temkin
   type. This suggests on the other hand in a first approximation
   an Elovich type kinetics for the redox reactions (iii).

Conclusions

     General Remarks. The wavefront analysis as presented here,
is limited only to the chemical relaxations and excludes any dissipation effects. The space relaxation, manifested as the change
of the propagating primary wavefront itself, is met with and corresponds to the initial steady state or an equilibrium state of
the system. On the other hand, the time relaxation manifests the
changes in the system, caused by the primary wavefront.
     From a thermodynamic standpoint, the space relaxation cor-
responds to the initial entropy of the system, i.e. to the initial entropy state on the catalyst surface, attained by establishing equilibrium between the ambient fluid and the catalyst
surface.
     The time relaxation corresponds to the change in the initial
entropy state. As it has been established by one of the present
authors (6), it is possible to derive generalized thermodynamic
forces and fluxes in the sense of Onsager's theory and to study
e.g. stability problems of tubular reactors.
     All the elements of wavefront analysis have the common cha-
racteristic that they depend only and in a straight forward way
on the initial steady state (for a nonlinear relaxation and/or
nonlinear wave propagation also on the amplitude of the disturbance). That means, they are directly accessible for kinetic analysis. It is also possible to achieve a splitting of the initial
information content of the transient response and make feasible
a direct measurement of the initial rate of the entropy change.
The space relaxation is the integral effect of the initial rate
of the chemical conversion, the initial time gradient of the time
relaxation is the integral effect of the initial acceleration of
the chemical conversion. The residence time of the wavefront
(dead time of the system) is the integral effect of the convective accumulative terms. For a lumped system all three effects
have been implicated in the initial time gradient of transient

response (see Table I). It seems therefore that the additional degree of freedom, introduced by the propagating wavefront, makes the new technique an extension to conventional dynamic techniques.

Specific Remarks. The established dependence of the microkinetics on the oxidation state of the catalyst make clear that a) results of kinetic investigations at lower temperatures are different in respect to the mechanistic scheme from those obtained at higher temperatures, b) in a distributed catalytic system in the steady state a distribution of the catalytic steps is possible as a direct consequence of the ambient gas concentration profile and the axial temperature distribution; in an extreme situation it is conceivable that at the reactor inlet, another mechanism dominates as at the reactor exit. These two facts can perhaps explain some contradictory results about the same reaction scheme which have been reported in the past by different authors. As stated recently by Boreskov (19) in a review paper, this conclusion holds true for the most catalytic systems under the technical operating conditions.
It is also clear that for a complete description of a catalytic system, as investigated here, one needs the complete functional describing the microkinetics in dependence on the energetic states of the catalyst surface.

The conclusions, established here, regarding the dependence of the overall activity as well as of the individual microkinetic parameters on the energetic states of the catalyst surface are in agreement with the fact that only the coexistance of oxides of Zn, Cu and Cr make it possible to have a good catalyst. That means a catalyst at intermediate oxidation state is an active one, whereas the oxides by themselves are not good catalysts (see the review paper (20)). Investigations on the dependence of the rate constants and activation energy on the chemical composition of the catalyst (see (20)) are similar to the reported ones here, where the change in the chemical composition is implied by the exposure of the catalyst on different ambient gas mixtures.

Especially for the low temperature water gas shift reaction the mechanistic scheme, proposed here, seems to correspond to the three different adsorbed oxygen species, proposed by Kobayashi (13) for the ethylene oxidation on silver, whereas the importance of some surface complexes of $CO - H_2O$ type has been revealed (14) by analysing steady state data.

The significance of the coupling of micro- and macrorelaxations for resonance phenomena observed in catalytic systems under forced periodic operation (cycling) (15) implies that the wavefront analysis of transients of this kind can eventually suggest a more effective strategy in seeking the optimal conversion and selectivity. Finally the existence of certain surface structures and complexes could be established, if the transients of the surface intermediates will be followed e.g. by infrared spectroscopy (see e.g. (16, 17)).

Table I

Elements of response	Wavefront analysis	Conventional transient response techniques
Propagation speed	$\ll$ Convective accumulative effects	_____
Time relaxation	$\ll$ Entropy change of the system in Onsager's sense	$\ll$ Convective accumulative effects and initial entropy of the system and entropy change of the system
Space relaxation	$\ll$ Initial entropy of the system	_____

Appendix

We regard here a catalytic reaction with the following stoichiometric formulation:

$$A_1 + A_2 \rightleftharpoons A_3 + A_4$$

If one stimulates an individual reaction step, in order to perform a qualitative or quantitative analysis, one uses a step inlet disturbance in one of the species keeping all other constant.
It is obviously that the ideal situation exists, when the time
constant of the measuring device, of the step signal itself (i.e.
the signal has to be "sufficiently ideal") and the residence time
of the wavefront in the system τ_0 stand in a reasonable relationship to the space and time relaxation. The transients of the
disturbed species consist then of clearly separated space and
time relaxation and those of the undisturbed species depend on
whether there exists for these species a measurable internal
storage or not. If not, the transients of the undisturbed species
consist of the step disturbance part (space relaxation) and of
the time relaxation part, otherwise they consist only of the time
relaxation part. When e.g. under transient conditions, a product
desorption step is kinetically controlling, its transient caused
by a step change in the reactant concentrations consists only of
the time relaxation part (see Figure 4b). If the same step is not
kinetically controlling the situation of Figure 4a is obtained.
The time relaxation in Figure 4b, i.e. in a simple situation, is
directly connected to the internal storage and the interchange
rate between the phases. When a measurable internal storage exists but the interchange is not kinetically controlled, the transients have the characteristic form of Figure 4c, where the over-
shooting type effects and the shift in the transient responses is
caused by the equilibrium between the fluid and solid phase, being typical for chromatographic reactors (8).
For the nonideal situation which does not fulfill the above mentioned requirements, the wavefront analysis fails to give a detailed information about the mechanistic scheme. But in any case,
the relaxations, measured at the wavefront, correspond at least
to the stimulated limiting step.

List of Symbols

E	activation energy
f_i	relaxation functions like chemical rate, heat generation etc.
f_{ij}	$\partial f_i / \partial y_j$
k_i	kinetic constants
m_i	eigenvalues equal to the inverse time constants of the space relaxation caused by the reaction system (i)——(iii)

n   reaction order
p_i partial pressure of the component i in the fluid phase time
t   time
u   average fluid velocity in the convective direction
y_i state variables like concentration in fluid and solid
         phase, temperature etc.
y_1 state variable with the largest propagation speed
Z,z   length of the catalyst bed, space coordinate
α_o propagation line of the wavefront
θ_i solid phase concentration
λ_1 accumulation parameter
τ_i Time constants
τ_o residence time of the wavefront in the system (hydrodynamic
         time constant)
ψ p_{H_2O}/p_{H_2} in the pretreatment period

Indices

eq   equilibrium
max   maximum
c   referred to fluid concentration
θ referred to solid concentration
time
relx   referred to time relaxation
space
relx   referred to space relaxation

Literature Cited

1 Temkin, M.I. Kinet. Catal. 1976, 17, 945
2 Temkin, M.I.; Nakhmanovich, M.L.; Morozov, N.M. Kinet. Catal.
   1961, 2 650
3 Fiolitakis, E.; Hofmann, H. Dependence of the Kinetics of the
   Low Temperature Watergas Shift Reaction on the Catalyst Oxygen Activity, to be published in Journal of Catalysis
4 Voorhoeve, R.J.H. ACS Meeting, Catalysis under Transient Conditions, Las Vegas, Nov. Aug. 25-29, 1980
5 Fiolitakis, E.; Hofmann, H. Lecture Notes of "Summer School
   on Modeling of Dynamic Systems Based on Experimental Data
   with Chemical Engineering Applications", Bad Honeff, W.Ger-
   many, Aug. 1980, Ed. A. Pethö - submitted for publication in
   Catal.Rev.-Sci.Eng.

6   Fiolitakis, E., Some Aspects on the Entropy Change in Onsager's Sense for Irreversible Chemical Processes, to be published

7   Fiolitakis, E.; Hoffmann, U.; Hofmann, H. Chem.Engng.Sci. 1980, 35, 1021

8   Schweich, D.; Villermaux, J; Sardin, M. AIChE J. 1980, 26, 477

9   Fiolitakis, E.; Hofmann, H. Chem.Ing.Techn. 1979, 51, 800

10 Aris, R.; Amundson, N. "First order partial differential equations with applications", Prentice Hall, Englewood Cliffs 1973

11 Van Herwijnen, T; de Jong, W.A. Jl. Catal. 1974, 34, 209

12 Kaleńczuk, R.J.; Fiolitakis, E.; Hofmann, H. Chem.Ing.Techn. 1980, 52, 966

13 Kobayashi, M. ACS Meeting, Catalysis under Transient Conditions, Las Vegas, Nev. Aug. 25-29, 1980

14 Van Herwijnen, T.; de Jong, W.A. Jl. Catal. 1980, 63, 83

15 Jain, A.K. ACS Meeting, Catalysis under Transient Conditions, Las Vegas, Nev. Aug. 25-29, 1980

16 Hegedus, L.L. ACS Meeting, Catalysis under Transient Conditions, Las Vegas, Nev. Aug. 29-29, 1980

17 Bell, A.T. ACS Meeting, Catalysis under Transient Conditions, Las Vegas, Nev. Aug. 25-29, 1980

18 Keil, W.; Wicke, E. Preprints of Meeting on "Kinetics of physicochemical oscillations", Aachen Sept. 19-22, 1979, vol. I, 172

19 Boreskov, G.K. Kinet. Catal. 1980, 21, 1

20 Newsome, D.S. Catal. Rev.-Sci. Eng. 1980, 21, 275

21 Fiolitakis, E.; Hoffmann, U.; Hofmann, H. Chem.Engng.Sci. 1979, 34, 677

22 Fiolitakis, E; Hoffmann, U.; Hofmann, H. Chem.Engng.Sci. 1979, 34, 685

RECEIVED July 28, 1981.

INDEX

Jacket design by Kathleen Schaner.
Production by Janet Shoff and Karen Gray.

Elements typeset by Service Composition Co., Baltimore, MD.
Printed and bound by Maple Press Co., York, PA.